D0258102

C.A. HOOKER taught philosophy and environmental engineering at the University of Western Ontario. He currently holds the chair of philosophy at the University of Newcastle, Australia. R. MacDONALD teaches environmental studies at York University. R. VAN HULST teaches biology at Bishop's University, Lennoxville, Quebec. PETER VICTOR is a research consultant with Victor and Burrell. He teaches environmental studies at York University, part-time, and is a research fellow at the Institute of Environmental Studies at the University of Toronto.

As the supply/cost crunch tightens, issues related to energy become increasingly compelling. This is a guide for the general public to the fossil fuel crisis facing Canada, and Ontario in particular. It is also about other long-term matters of greater importance: the economic, socio-political, and cultural consequences of the choices which now have to be made, primarily by governments.

The authors argue that energy policy is social policy. Therefore our ideas about the kind of society we want must be a governing consideration in working out a policy to take Canada through the energy crisis.

The four writers bring to bear on the problem the perspectives of engineering, philosophy, environmental studies, and economics. The result is a balanced guide for the continuing debate on the adaptation of society to the imperatives of energy.

C.A. HOOKER
R. MACDONALD
R. VAN HULST
P. VICTOR

Energy and
the Quality of Life:
Understanding
Energy Policy

UNIVERSITY OF TORONTO PRESS
TORONTO BUFFALO LONDON

© University of Toronto Press 1981
Toronto Buffalo London
Printed in Canada

ISBN 0-8020-5514-1 (cloth)
    0-8020-6410-8 (paper)

Canadian Cataloguing in Publication Data

Main entry under title:
  Energy and the quality of life
  Bibliography: p.
  ISBN 0-8020-5514-1 bd. ISBN 0-8020-6410-8 pa.

  1. Energy policy – Canada. 2. Energy policy –
  Ontario. I. Hooker, Clifford A. II. Title.
  HD9502.C32E53   333.79′0971   C80-094489-5

This book has been published with the assistance of the Canada Council and
Ontario Arts Council under their block grant programs.

Contents

Preface vii

PART I:
ENERGY AND THE QUALITY OF LIFE
1 A framework for energy policy-making 3

PART II:
A CRITIQUE OF CONVENTIONAL ENERGY POLICY
2 Energy dilemma and policy response in Canada 23
3 Energy dilemma and policy response in Ontario 47
4 Nuclear energy policy in Ontario 64
5 Energy policy as social policy 84

PART III:
AN ALTERNATIVE ENERGY POLICY
6 An alternative energy strategy 99
7 Energy conservation 102
8 Technology for an alternative energy strategy 130

PART IV:
FROM ENERGY POLICY TO PUBLIC POLICY
9 Institutional structure and energy policy 169
10 Understanding energy policy 202

Notes 209
Reader's guide to the bibliography 251
Bibliography 255
Contacts in the alternative energy field 281

'The energy policies adopted during the current decade will determine the range of social relationships a society will be able to enjoy by the year 2000.'
Ivan Illich

Preface

Our work has evolved over a period of five years, from an initial encounter at a conference on politics and ecology to the present final product. We had two profound problems to overcome: how to orchestrate four very individual individuals; and how to resist the temptation to respond to every crisis and event the turbulent energy field has presented over the past five years.

As to the latter problem, we believe we now have a book that strikes a balance between eternal verities and momentary events, focusing on the important issues in this field facing people everywhere and not just in Canada – which is what we always told ourselves we wanted. The energy situation is almost certainly going to become much more, not less, turbulent in the 1980s, and people will only avoid the heavy penalties of poor or chaotic choices if they first reflect clearly on what is at stake.

As to the former problem of orchestration, all four of us jointly take responsibility for the entire book. Readers who know about our backgrounds will guess correctly that C.A.H. and R.V.H. took the initiative in preparing material for Parts I and IV, and R.M. and P.V. did likewise for Parts II and III. Subsequently, however, each of the contributions has been argued over, dissected, and rearranged by the group of us, an exhausting but rewarding process that has lent a cheerful unanimity to the result. This unanimity should be understood to cover the importance of the issues raised, the perspective from which they are raised, and the general directions in which responses are sought. Unanimity should not be presumed on all matters. In particular, being the individual valuers we are, we differ among ourselves on the emphasis placed on various issues and responses – for example, on radiation damage versus damage from conventional energy technologies, on the socially controlled use of economic markets versus non-market alternatives, on the value of investing in self-sustaining 'arks' versus providing mass-produced alternative energy technologies. But these are, after all, matters of emphasis in an agreed framework. And in any case they are matters to be decided democratically by society through experimentation and debate. We simply signal our personal differences here.

Ironically, our most serious disagreement came over the importance of editorial work as C.A.H. wanted to accord it less value than did the others. It was resolved in the following way: C.A.H., whose original initiative got the project started, acted as conductor of the orchestra, rewrote much of the manuscript to achieve greater stylistic coherence, prepared the figures, tables, diagrams, footnotes, and bibliography, and generally oversaw the book's safe production. The book itself is our reward, and that is as equally distributed as were the contributions to the content.

The book went through several drafts, each a wild array of pen and pencil over type (except the first, which was a wild array of pen and pencil). These successive drafts were patiently and accurately transformed into would-be manuscripts by Ms Penny Lister, who grew to have as many parental feelings about the book as we did ourselves. Certainly she laboured as hard in delivery as any of the rest of us. Penny also drew the figures and applied her aesthetic judgment to the internal arrangement of both figures and tables. To whom be thanks eternal.

Many other people made not inconsiderable contributions; in particular Mrs Bev Hughes typed the footnotes and bibliography and made last-minute manuscript corrections under great pressure of time – while Penny delivered and enjoyed her real baby.

Work on this book was completed while C.A.H. and R.V.H. were supported by a Social Sciences and Humanities Research Council grant under which many of the ideas were developed. The grant was interrupted for six months to take up a research contract with the Ontario Royal Commission on Electric Power Planning, issuing in a report *Institutions, Counter-Institutions and the Conceptual Framework of Energy Policy Making in Ontario* which proved useful in the preparation of Part IV of this book. C.A.H. and R.V.H. are grateful to both Council and the Commission for their support. R.M. would like to acknowledge the generous support of the Faculty of Environmental Studies, York University, throughout the preparation of the book. Publication has been assisted by the Canada Council and Ontario Arts Councils through their block grant programs.

Finally, we should like to acknowledge all those people who gave so unstintingly and unselfishly of their time and energy to help us in the preparation. Special mention may be made of Peter Middleton, Middleton Associates, who was with us in the beginning and provided excellent advice on the focus and approach of our work, offered much encouragement and loaned his files and coffeemaker; we regret that circumstances forced his withdrawal. Our particular thanks, too, to the staff of the Royal Commission on Electric Power Planning. To attempt to list the many people from various private organizations and government agencies would be impossible and we fear injustice in omitting some, so we offer a blanket paean of praise and remind ourselves again that what ultimately makes for good things socially is the goodwill of people working informally to create them.

PART I

ENERGY AND THE QUALITY OF LIFE

1
A framework for
energy policy-making

The plain fact is that while Canada is almost wholly dependent for energy upon its fossil fuel deposits, principally oil and gas, supplies of them are dwindling and the costs of finding and extracting them are rising, all very rapidly. This, in a nutshell, is the energy 'crisis.' This book analyses that crisis and our future energy options.

But it is also about something still more important: the potential economic, social-political, and cultural consequences of the energy choices that must now be made. The possible results of our choices have tended to remain buried and neglected in the welter of technical detail; yet we believe that they are of more importance in the longer run than the technical considerations, and that evaluation of them is the essence of responsible choice of future energy policy. While we focus in this study on energy policy in Canada and the province of Ontario in particular, we have also aimed to write about the universal human problems in energy policy-making. To grasp the wider dimensions of energy policy and their complex interactions and the proper evaluation of energy data it is necessary to develop an appreciation of the nature and role of energy in our lives.

ENERGY AND REALITY

To live is to transform energy. Energy reaches earth from the sun in the form of radiant sunlight. A small fraction of it, roughly 1 per cent, is captured in photosynthesis and transformed into chemical form as proteins, cellulose, and other structures in plants on land and plankton in the ocean. From this base every living thing in turn derives its sustenance, by first consuming this chemical energy as food and then transforming it into body structure, heat, and movement. Cattle eat grass; humans eat cattle. The energy transformations in the biosphere end only when the original solar energy is finally reradiated back into space as low-temperature heat. This happens directly in a bush fire or when plant eaters burn the chemical energy for body heat. The energy of most movement, too, is quickly transformed into heat via friction. The process is delayed only slightly if the energy is first transformed down a food chain from prey to predator; for example, from grass to field mouse to snake to hawk. Finally, all body wastes, and bodies themselves at death, yield their remaining chemical energy as heat as bacterial action decays them, breaking down their chemical structures for use in the next cycle of life.

The energy flow may be indefinitely delayed if the chemical energy remains trapped in, for example, coal and oil deposits. In exploiting such deposits, we succeed only in completing the transition, finally releasing the solar energy to space as heat. Indeed, if we move back to the grand cosmic scale of the birth and death of stars, galaxies, and perhaps the universe itself, we can view even the energies stored in the earth's centre, and in radioactive nuclei such as those of uranium as part of a larger cycle of energy flow and transformation. Modern physics increasingly supports the view that, ultimately, everything is energy – energy in flow or in transformation. As Einstein's theory states, matter is a form of energy. Seemingly substantial forms are but relatively stable patterns in the universal energy flux.

Thus what we understand about energy and how we subsequently relate our choices to its flow and transformation are the most basic issues facing us. As yet our species is young by evolutionary standards, our understanding is very limited, and our choices are constrained by our relation to this planet's natural environment and by our handling of the special additional energy flows our industrial activity has introduced. These choices are, however, basic to the present quality of life and to providing a future where understanding and choice will widen, rather than wither.

The whole scheme of living things is first and foremost a system of energy transformers that together comprise the wonderfully complex web of energy flows and transformations that is the biosphere. Every living species has its place in this web; it 'makes its living' by acquiring, transforming, and passing on the basic energy at some special location within the weave.

The earth's climate, the backdrop to this dynamic drama, is also largely controlled by the same flow of solar energy. About 40 percent of the energy flow reaching the earth from the sun is absorbed by air, land, and oceans and controls the climate. This energy drives the great ocean currents and wind systems of the planet, it evaporates the water to form cloud and rain, and it heats the land and waters. Together with physical geography, these great energy flows shape the planet's climate. The world's great river systems, the basic formation of soil from weathering of rock, and winter snow, spring storms, and the tropical monsoon are all consequences of the solar energy flow.

To our untutored perceptions, the earth often presents an appearance of solidity, stability, and vastness, an ageless superhuman framework in which we pass our daily round. In reality it is better viewed as a small ball bathed in a vibrant wreath of energy that forms a thin web of intricately interwoven energy flows and transformations. We experience this energy web as living. It is dynamic, not static. Everywhere there is the evidence of change: the record of evolutionary development, the human devastations of landscapes, and the passing of a summer storm or a human life.

In the last three hundred years, an instant of historical time, we have introduced a small addition to this energy web with our industrial activity. Still, the natural structure of energy flows forms the fundamental framework for life as we know it. A basic focus of energy policy is the choice of our relationship to this enveloping web of life. Thus energy policy concerns the most basic questions of the organization of human society in relation to the natural environment and to one another as living creatures.

As for animals long before us, the physical secret of our survival and the success of our civilizations rest with our ability to adapt the natural environment so as to capture a larger share of the available energy. Early in our history man built wood fires and domesticated plants and animals, making more solar-based chemical energy available per capita. Then man learned to utilize the wind and waterpower in ships and mills, making solar-based non-chemical energy available. And now we have successively utilized coal, oil, gas, and uranium – all but the last solar-based chemical energy. A second major concern of energy policy is, therefore, with the choice of additional energy sources and with the selection of how their flows and transformations will mesh with the natural energy web. These choices will determine no less than the basic form of our urban and industrial society.

Every machine process, every computer or human communication, every surplus 'man-hour' that can be wrested from basic survival effort and devoted to culture and civilization is purchased with energy. Energy availability is a necessary condition of a high quality of life. Energy provides choices. Without energy we have no choices, and human freedom and dignity will be diminished. This is not to say, though, that increasing energy consumption is always better. Energy may also be abused or used to provide poor choices or to limit them. In the past, for example, new energies have been used for military destruction and political repression on unprecedented scales.

The appropriation of energy also imposes its own constraints. A society has to be willing to organize its structure around the acquisition and distribution of energy. And as the energy supply increases, so too does the social dependence on that energy. The costs associated with energy technologies – economic, environmental, and sociopolitical – also mount with increasing energy dependence.

In the past, the costs associated with increased energy availability largely derived from damage to the environment or abuse of the choices this increase made possible. The coal-based air pollution of industrial cities in nineteenth- and twentieth-century England and the oil pollution of oceans today are examples of environmental costs. The increase in the lethal nature of warfare following the invention of gunpowder and the atomic bomb are examples of abuse of choice. Historically, however, these costs have usually been counterbalanced by the substantial increase in individual choice among consumer products, home heating and cooling systems, rapid transportation, and so on. However, much of the time the poor and disenfranchised have borne a disproportionate share of the costs, while the benefits have been largely enjoyed by the wealthy who have dominated in making choices about energy.

At the present time the increase in individual choice is tempered by the rising social and environmental costs associated with energy systems. Our societies also have to pay all the associated costs of continuing to demand large energy subsidies. While in the past the poor have lived in the air pollution of mines and smokestacks, even the millionaire must now suffer from congested roads and the risk of nuclear accident. These costs have become dramatically more apparent in the last decade. Some of them arise from the very large and rapidly increasing economic costs to develop energy supply technology, costs increasing to the point that they change the balance of economic life in important ways; some from mounting environmental costs associated both with energy extraction technology, as at James Bay, and with energy transformation technology, as with radiation risks from nuclear-electric plants; and some from subtle, but ultimately crucial, social and political costs centring around national and international domination by centralized, relatively inflexible energy corporations and bureaucracies, pursuing policies that involve social commitments for generations into the future.

Industrial societies now find themselves in unprecedented circumstances. The immediate cause of the air of crisis currently attached to energy is the impending failure of the present fossil fuel sources on which humans are most dependent. Our societies are for the first time in history annually consuming significant fractions of the total accessible planetary supplies of the most attractive fossil fuels – oil and gas. Their relative exhaustion looms on the horizon. Yet because these very energy sources are consumed in such large quantities, our societies also have an unprecedented reliance on them. Our entire way of life at present revolves around their flow. Energy is like a drug; there can be addiction, even in the face of mounting costs. Withdrawal symptoms following denial of the drug can be very distressing.

In terms of the earlier analogy of the energy web of life, the internal microcosm of our industrial societies is one vast web of coal, oil, gas, and electric energy flows and transformations. This web extends out into the natural environment, having created, for example, energy-intensive industrial agriculture and silviculture. A large part of the current debate over energy policy concerns how society can ensure adequate replacements for present energy dependencies.

That the choice of energy flow also affects all the more obvious aspects of social life – for example, industrial structure, urban structure, and employment – should be quite clear. Indeed, once the full significance of the energy web has been grasped, it should come as no surprise that energy policy affects every aspect of human life, even the most apparently 'non-physical' factors such as culture and politics. If, for example, the available energy flow is diverted primarily into the production of consumer goods, one kind of culture will be reinforced; if, however, the energy flow is directed towards the provision of public services, community resources, and decentralized urban structure, quite a different culture will result. Energy is a fundamental basis for choice, and control of energy is the control of choice. If the supply of energy falls, we shall either have to take harsh political measures to maintain existing social inequalities of wealth or become a much more sharing, egalitarian society. If the supply of energy is to remain at its present level, or to rise, we shall either have to take harsh political measures to ensure that the costs are born by those who benefit – and people do not benefit equally – or be willing to subject our choices of energy use to searching social scrutiny.

Thus the ubiquitous nature of energy choices opens them up to public decision-making in a very important way. If energy is basic to access to choice, it follows that the choice among energy alternatives – that is, energy policy – is itself a basic social choice. It is one of our distinctively human capacities to have foresight and a distinctive human ability to take responsible action to bring about the future in accord with the highest human values. Our energy circumstances call for nothing less than this from us.

Fortunately there is considerable scope for choice in our current circumstances. There is first the choice to use energy more efficiently, to waste less, and to utilize the transformations in the energy web as effectively as possible. This involves a complex set of choices; for example, concerning insulation, waste heat recovery, judiciously combining processes, and ultimately about matching the form of energy to its use. The immediate costs of acquiring energy are thereby reduced. More important, this choice allows time to make the larger decisions in a proper manner, so that we do not rush into them out of fear, ignorance, or political inadequacy.

A larger choice always available to us is between acquiring ever larger energy flows and adopting a way of life that is both less energy-intensive, more service-oriented, and communal. This is not, of course, a single simple choice but a general label for a vast array of choices about work sites in relation to home locations, kinds of transportation, forms of housing, and so on. To make such choices brings into play the full gamut of human values; to make them in a fair and deliberate manner will challenge all the combined expertise in democratic process.

Fortunately, too, we do not need to make a simple-minded choice between one massive energy technology and none at all. History may indicate otherwise, but there are many different energy technologies among which to choose, at least in the next fifty-year period in which our energy problems will come to a head. Each of these technologies fits into some ways of life, fits with some decisions about the design of our engergy webs, better than others. Thus we have additional leeway to reinforce those sets of human social values we prefer by choosing a mix of energy technologies that best fits with our chosen social form.

It is fortunate for us that all these choices are still open. They will not be nearly so open in twenty-five years. By failing to make any deliberate choices now, we shall likely have locked ourselves into inadequate energy supplies and into a group of unpalatable energy technologies. We have a unique opportunity now to decide the future. We can choose a future in which there is maximal flexibility of choice of energy policy for at least the next two succeeding generations.

The choices are, of course, extremely complex and multifaceted. This book explores the nature of the choices by focusing on the circumstances in a particular country, Canada, and a particular region within it, the province of Ontario. We hope thereby to provide a framework for examining the choices not only for the people of Ontario, or Canada, but also for societies everywhere by showing the universal structure in the particular instance. A more detailed account of two aspects of this framework is essential to understanding the analysis of energy planning that follows. Thus a thermodynamic account of energy quality that is essential to evaluating alternative energy technologies and conservation opportunities is outlined. Second, the nature and scope of an adequate energy policy and its implications for government responsibility are discussed. Within this framework we can approach the question of the content of an adequate energy policy.

ENERGY DEMAND, THERMODYNAMIC EFFICIENCY,
AND ENERGY ACCOUNTING

Energy is only useful to us because of its ability to do useful work when harnessed by technology. Given the rising costs associated with obtaining energy, it makes good sense to increase the efficiency with which we use energy to do work. In attempting to use energy more efficiently, the science of thermodynamics provides important insights. This scientific discipline has been compressed into two succinct statements, often called the First and Second Laws of Thermodynamics.

The First Law states that in any process in which useful work is done, energy is neither created nor destroyed but only changed from one form to another. In many applications in which energy is converted from one form to another, and in the process useful work accomplished, energy in the form of heat is either involved directly or produced as a byproduct.

The Second Law states that it is impossible to convert heat energy into mechanical energy with 100 per cent efficiency. In any process using energy, some portion of the ability of that energy to do work is always lost, for in all processes involving the use of energy, heat energy is either involved directly in the process or is produced as a byproduct in the form of friction or noise or vibration, all of which are quickly absorbed by the surroundings and show up as heat. This heat energy is always at best only partially available to do useful mechanical work.

In terms of our earlier analogy of the web of energy flows and transformations, the web can be pictured arranged vertically with energy entering at the top and cascading down along the various pathways. Each branchpoint in the web and each transformation along the way decreases the usefulness of the energy, with some energy always ejected at the bottom of the web as low-grade heat. Eventually all the energy reaches the bottom and is ejected as low-grade heat.

The First Law of Thermodynamics tells us that energy is never used up, while the Second Law tells us that its important quality of being able to do useful work is used up. Thus an energy crisis is really a Second Law thermodynamic crisis. We are not running out of energy, which is impossible by virtue of the First Law; we are facing shortages in the ready availability of those conventional forms of energy capable of doing large amounts of useful work.

One way of maximizing efficiency in the use of energy in an industrial economy is through the process of 'thermodynamic matching' in which forms of energy that possess a large capacity to do useful work are matched with end-uses demanding that capacity. The importance of this criterion is that as high-grade energy forms are both increasingly costly and increasingly scarce, we need to use them as effectively as possible.

As an example consider certain industrial processes, such as metal smelting, that require very high temperatures. The ability of electricity, oil, or natural gas to achieve high temperatures makes them appropriate for industrial uses. The quality of the energy form is 'matched' to the thermodynamic requirement of the energy end-use. By contrast, in residential space heating, air or water temperatures of about 55 degrees Centigrade are required. Electricity is a high-grade form of energy capable of producing temperatures of several thousands of degrees. Thus to use electricity for household space heating would constitute a thermodynamic mismatch between this high-quality energy form and the end-use. In such an application, most of the capacity of electricity to accomplish useful work is wasted. In fact, this low-temperature heat requirement can be 'matched' by solar radiation at comparable economic cost and with lower environmental and social costs.

In many instances, then, much of the capacity of our expensive and depleting energy resources to do useful work is being wasted through a failure to match the energy form to the social application. Analysis of our national patterns of energy use reveals that there are three broad categories of end-use: one in which energy is required in the form of heat, mainly process steam and space heating; a second in which energy is required in the form of high energy-density liquids, principally for the transportation sector; and a third in which the particular characteristics of electricity are specifically required, mostly for lighting and electronic equipment.[1] The heat-requirement category is by far the largest, taking up as much as 50 per cent of all end-use energy. About 80 per cent of the heat needed requires temperatures below 140 degrees Centigrade. Since currently fossil fuels and electricity are largely used to provide this energy, there is a large potential for energy savings without forgoing any present uses.[2]

Of course thermodynamic matching only gives a first indication of potential energy savings. Whether the savings are 'real' needs to be determined by comparing energy alternatives for the situation throughout the entire web of energy flows and over the full lifetime of the equipment involved. It might turn out, for example, that construction of new equipment to take advantage of energy saving itself required more additional energy than would be saved, or that the equipment had to be replaced so often that, averaged over a long period, there would be no net energy saving. However, the large-scale thermodynamic mismatches that operate in contemporary industrial societies can certainly be reduced in an energy-efficient manner and indeed promise potentially large energy savings for industrialized nations.

Similarly, not every energy saving is economically attractive. Whether an increase in Second Law efficiency is economically advantageous depends upon the availability of appropriate technology and its cost, the legal and social consequences of its operation, the economic productivity of this use for capital in comparison with competing uses, the prevalence of monopoly pricing in the energy sectors, and so on. More subtly, it depends on large-scale social decisions about desirable energy forms and technologies. For example, electricity could be costed so as to entice industries to use it for space heating despite the resulting thermodynamic mismatch; conversely, direct solar space heating may be made economically attractive through a programme of government incentives for mass development. Here, as in other areas, individual decisions about investment mesh in a complex manner with social decisions about preferred ways of life and national interests. What is economic emerges out of this interaction. Neither thermodynamic efficiency nor simple economic advantage can be used as a final arbiter of energy choices; rather they are two ingredients in a complex process of individual and social decision-making. This book is about that process.

Throughout this book quantities of various energy forms are discussed in terms of a common unit of measurement, the British Thermal Unit (BTU). The BTU is the amount of heat energy needed to raise the temperature of one pound of water by one Fahrenheit degree. In more meaningful terms, perhaps, it takes about 75 million BTU to heat the average house for a year, about the same to run the average car, and about 7,000 BTU daily (2 million BTU annually) to maintain body functions.

Energy can be converted in various ways from one form to another. When other forms of energy such as coal, oil, or electricity are compared, their energy content is expressed in terms of the amount of heat liberated when these forms of energy are converted into heat under optimum conditions – for example, through chemical combustion or in electrical resistance heaters. Conversion rates used in this book are as follows: oil, 5.8×10^6 BTU per barrel; natural gas, 1.0×10^6 BTU per thousand cubic feet; coal, 26.2×10^6

BTU per short ton; and electricity, 3,412 BTU per kilowatt-hour. Note, however, that expressing all forms of energy, regardless of quality, in terms of their BTU heat content, while convenient when comparing differing energy flows, entirely misses the very important qualitative aspect of energy use: the capacity of a particular energy form to accomplish useful work.

In energy accounting it is important to separate and keep track of the demand for both primary and secondary energy. Secondary energy is that energy used for direct end-use demand, such as that energy used in lights or machines. Primary energy is that energy used in the overall energy system. Thus primary energy is equal to secondary energy plus the internal energy consumed by the energy production sector.

Demand for primary and secondary energy may differ widely, depending on the energy production technology employed. For example, suppose 3,412 BTU of high-temperature heat is required at a metal smelter. One way to provide that energy would be to burn oil directly, say at 90 per cent efficiency in an oil-fired furnace; then corresponding to a secondary energy demand of 3,412 BTU at the smelter there is a primary demand of about 4200 BTU of oil, which includes a loss of about 400 BTU in transmission. An alternative way of providing that energy would be to burn oil in an electrical generating plant, at about 35 per cent efficiency, and use the electricity to provide heat to the smelter. In this case the smelter's secondary demand of 3,412 BTU amounts to 1 kWh. But now the primary energy demand is 10,000 BTU plus the roughly 10 per cent more (1,000 BTU) consumed internally elsewhere in the energy production and distribution technology. That is, about 11,000 BTU of oil are consumed to perform the same task. In both cases in this example the same amount of usable energy is delivered at the smelter; yet the primary demand on resources differs greatly. This is why we have insisted on valuing electricity at 3,412 BTU/kWh, which is its BTU equivalent as secondary energy demand. The alternative figure often used – 10,000 BTU/kWh – is only appropriate to primary energy demand and only in the context of substituting electricity generated by fossil fuels in plants with steam-driven turbines for electricity generated by either nuclear or hydraulic energy. When represented at its end-use value of 3,412 BTU/kWh, electricity is seen in proper perspective as a much less important component of total energy demand than would appear when valued at 10,000 BTU/kWh.

Primary demand is important for two reasons. First, it reflects production efficiency (or inefficiency); when energy is supplied inefficiently from scarce resources such as oil, it is important to understand the resource implications of primary energy demand. Second, the difference between primary and secondary energy demand is a measure of waste heat ejected into the environment. When such ejections are large they may disrupt the local ecology; alternatively they may be used as a basis for some industry. Either way, it is important to keep track of them.

When an energy resource is renewable, production efficiency becomes a less important consideration, except that improvements in efficiency will presumably often be economically desirable. Moreover, when the energy is already widely spatially dispersed, as is solar energy, waste heat is also a less important factor. In the case of renewables, therefore, the focus of attention should be on secondary demand and the socioeconomic implications of meeting it or shaping it. This is where we place the emphasis when we discuss these energy technologies in chapters 8 and 9.

In fact, discussion of energy policy should always begin with secondary energy demand for two basic reasons. First, secondary energy demand concerns the immediate end-use demand at the point of social application. This is the appropriate focus of energy policy decisions dealing with the social characteristics that determine end-use energy demand; for example, gasoline demand by motorists and electrical demand by households. Second, primary demand can be calculated from secondary demand once the energy technology mix is specified. But choosing that mix is another fundamental dimension to energy policy. A third fundamental component of energy policy is, of course, consideration of the social and environmental consequences of primary energy resource exploitation.

Thermodynamic efficiency and energy accounting provide a useful framework for the evaluation of energy policies. They form the basis for understanding why Canada now faces rapidly rising primary energy demands, and they also provide a physical framework within which one can evaluate the difference between attempting to meet those primary demands with yet more primary energy exploitation technology (oil wells, etc.) and trying to cut secondary demand while improving efficiency of use and production. Of course, there is more to evaluating energy policy than consideration of an adequate physical framework. It is also necessary to have an adequate social, political, and ethical framework; we seek to offer pertinent elements of this as well.

THE NEED FOR A SYSTEMATIC ENERGY POLICY

Our discussion has emphasized that energy policy is fundamental to all social policy-making and that many of the most important areas of government policy responsibility are directly affected by the choice of energy policy. For example, the relative availabilities and costs of coal, oil, and electricity will have a profound impact on whether transportation policy promotes, say, motor vehicles, steam trains, or trolley cars. There will be similar effects on methods and costs of home heating. In addition, the choice of energy production technology has an important impact on the structure of industry and employment. For example, consider the differences in jobs and industries among societies in which all energy comes from coal mining via steam engines and steam turbines, from oil refining and combustion, from nuclear electric energy, or from solar radiation. Of course, real energy options will involve mixes of these and other energy production technologies; the important point is the way the choice of these technologies will affect our way of life. How these technologies are financed and who owns them can make a very important difference to economic performance and to the distribution of wealth and power within society. Thus the choice of an energy policy is a matter of great social and political moment. It amounts to the choice of the most basic structural principle in society, and every facet of the society feels the effects of that choice.

Conversely, there are many areas of social policy where policy implementation has an important impact on the energy economy. Therefore, the government has as many tools it can mobilize behind the implementation of an energy policy as there are interacting consequences. If it is true, for example, that energy policy will affect the price of housing in various ways – for example, through heating technology costs – and therefore affect the choice of housing policy, it is equally true that building and insulation codes can be chosen so as to support a vigorous policy of energy conservation. If it is true that development of a centralized energy supply system largely encourages regional disparity, it is also equally true that a programme of regional development – including, for example, selective freight rates and tax incentives – can be deployed to favour the development of decentralized energy technologies.

From this perspective, it is clear that there are as many potential components to a public energy policy as there are ways open to government to influence socioeconomic development.[3] This is only the converse of insisting that energy policies cannot be developed in isolation from the remainder of social policy; rather, coherent integrated combinations of social policies need to be chosen. Choosing an energy policy is a fundamental part of choosing our future.

What are some of the more important policy tools available to government in pursuit of a given energy policy? The most obvious group are those economic incentives and disincentives that apply directly to energy itself and to energy supply industries. These include the direct pricing of the energy resource (e.g., setting oil and gas prices), the determination of rate structures for energy consumption (e.g., regressive or not), and taxation policy for energy supply industries (e.g., tax relief on profits reinvested in oil exploration or a gasoline retail tax). Less directly they include exploration permits, legislation permitting pipelines and setting criteria for their construction, federal financial backing of nuclear research (e.g., through Atomic Energy of Canada Limited), and provincial backing (e.g., Ontario Hydro's borrowing on foreign capital markets).

It is this set of policies which are commonly regarded as constituting energy policy. Hence, the current preoccupation with what the price of Canadian oil should be is thought by most observers to be a fundamental energy policy issue. Our view, and the reason we pay little attention to this matter in this book, is that the pressures for oil in Canada to be sold at or close to the world price will ultimately prove irresistible, given the major role played by the private sector in the oil market and Canada's increasing dependence on imported oil. This, we believe, underlies the stance taken by successive federal governments on the price of oil. In any case, the crux of the oil price issue is one of agreeing on a 'fair' profit for the oil companies and a proper distribution of tax and royalty revenues among different jurisdictions. These are as much issues of political and economic concern as they are of energy policy.

A second group of policy tools includes economic and legislative policies that directly affect the energy economy but are not applied to the energy production system. These include freight rates, fleet mileage standards for motor vehicles and trucks, loans and subsidies to transportation sectors (e.g., highways and domestic air travel), and transportation regulations (e.g., safety), all of which help to determine the mix of transportation modes that will be economically most attractive. Similarly, tax structures for fertilizers, pesticides, and heavy farm machinery help to determine the economic attractiveness of large-scale, energy-intensive farming in comparison with smaller scale, more labour-intensive farming. Again agricultural, health, and commercial regulations are an important influence on the mix of factory processed versus unprocessed foods that is economically attractive, and this in turn is an important factor in energy consumption in the food sector.[4] This second group of less direct policy measures also includes efficiency and pollution standards for industrial processes, insulation standards for private housing, heating efficiency standards for commercial buildings, and efficiency standards for electric motors and heating furnaces and the like.

A third group of policy measures indirectly important to energy policy are those that operate on a larger social scale and on a longer time base and concern the modification of socioeconomic structure and cultural values. The principal policy areas we have in mind are economic development, regional development, research and development, and education. The economic measures include industrial development incentives that would modify the industrial structure and commercial incentives and regulations that would modify, for example, the structure of the burgeoning service sector. The Canadian economy is oriented to heavy, energy-intensive, primary industry and to services rather than to manufacturing.[5] It is well known that primary industry is in general more energy- and capital-intensive than is manufacturing. Thus any shift in the industrial structure towards manufacturing would help both to reduce per capita energy demand and to create employment. Moreover, at least some primary industries can substantially improve their energy efficiency and can be offered incentives to do so. For example, the pulp and paper industry can be virtually self-sufficient in energy, and much less polluting as well, if it uses its own wastes for energy production; similarly, agriculture can be more energy self-sufficient.[6]

The growth of energy consumption – notably electricity – has, however, been largest in the services sector. This is not to say that the services sector is inherently more energy-intensive than the manufacturing sector. The fact that the large growth of electricity consumption in the services sector is a relatively recent phenomenon results from trends towards, for example, highrise office buildings, more energy-intensive heating and lighting practices, and large hospitals with high-technology medicine. As long as the present trends in the services sector persist, appreciable increases in energy demand will occur, the more so if that sector continues to expand at a disproportionate rate. Once again, government policy has a potentially large role to play. It can influence, at least to some extent, the relative rates at which the various sectors expand economically and, within the services sector, which services are encouraged and in what form. These policies will in turn have an important impact on the overall energy demand and use of the various sectors.

There are also many ways in which regional development is linked to energy policy. Most directly, decentralization of energy production may encourage the spread of decentralized energy technologies and make the exploitation of indigenous renewable energy resources more attractive (see chapter 9). Less directly, decentralized energy consumption tends to reduce demand for transportation, processed food, and other energy-intensive factors in the mix of social activities.[7]

By selectively choosing which technologies to support through the research and development stages, government can have a major impact on the mix of economically viable technologies available to a society at any given time. Many socially important technologies either would never be developed or would not be developed at the socially most appropriate time, and on a socially appropriate scale, if left entirely to market forces. As a rough rule, technologies whose social deployment requires long lead times, either in research or development, and/or initially requires large amounts of high risk capital, will fall into this category. Examples in the energy production sector are nuclear technology, tar sands extraction technology, and many waste recycling technologies. In the 1980s government support for nuclear and tar sands energy technology research and development stands out as a major component of the energy policies of the federal and some provincial governments.

Education is another important area in which government can affect energy policy. Until the 1960s environmental and resource issues generally were hardly touched upon in school curricula or in adult education programmes; energy resources in particular are an even more recent arrival on the scene. For more than half of the twentieth century, people have been 'educated' essentially without learning anything about the fundamental physical, biological, and social structure and function of their world and their own society. There does not even exist a clear, standardized form for presenting basic energy information to the public, and in many cases the information is simply lacking or publicly unavailable. To some extent, this situation is now being corrected in our schools, though in many instances the presentation is too shallow to be adequate to the importance of the subject; the adult education situation seems to be improving less quickly and effectively.

Education must move beyond this narrow technical confine to embrace the social, ethical, political, and cultural issues surrounding energy policy. If the presentation of basic facts is still in its infancy, the connection of these larger dimensions to energy policy has really not yet been made. What there is of public debate about energy policy is still largely carried on in narrowly technical and economic terms. Studies of actual institutional processes, of centralized versus decentralized models of decision-making, are again woefully lacking in relation to their importance in both school and adult education.

Similarly, the study of the relation of ethics and cultural values to public policy is underdeveloped, though perhaps to a lesser extent. Yet these larger issues are the essence of wise energy policy choices. Social morality and responsibility must ultimately be based on the basic ecological and social interdependencies among people and their environment. A politically responsible and responsive electorate is hardly to be expected if it is kept in ignorance and suffers the frustration of political impotence.

It emerges, then, that government has a very wide spectrum of policy areas that bear upon, and hence are bound up with, the formulation of public energy policy. The opportunity is wide, and so is the responsibility.

THE TASK AHEAD

We have now set out some conception of what a systematic energy policy might encompass and something of the physical framework in which it operates. In Part II this analysis is applied to the evaluation of existing government energy policy, at both federal and provincial levels. We argue that existing policy is inadequate by its own criteria in that it does not ensure adequacy of supply, on other criteria because it does not promote economic self-reliance, and by criteria for a reasonable quality of life. In Parts III and IV some alternative approaches to energy policy and policy-making are explored, showing that some mix of these is preferable because it is more likely to succeed where the existing set of policies failed. The challenge is thus presented both to government and to society to rethink present values and choices in the light of their likely future consequences.

Our criticisms are directed mainly towards government policy, as revealed in policy documents. We have almost entirely omitted from scrutiny the actions of the immense multinational corporations that dominate the commercial energy field. There are several reasons for this, the most important being that it is government that is ultimately responsible for energy policy and so it is government statements and actions that are the proper focus of scrutiny and counter-proposal. It is fair to say that as little as ten years ago there was barely an energy policy of any kind in Canada, let alone the sort of systematic energy policy we envisage. Instead energy resources were viewed simply as marketable commodities, in principle on a par with toothbrushes. Even seven years ago there was no federal Office of Energy Conservation and no Ontario Ministry of Energy. And while there is still no coherent public energy policy of the sort that is essential, there is a rapidly growing understanding of the need.

This is not to deny that the actions of private corporations have had a profound impact on energy policy today. But private energy corporations pursue much more limited and usually shorter term goals than are consistent with serving the public interest. These various goals are fragmented around the private exploitation of particular energy resources. We cannot therefore expect that any coherent energy policy in the public interest could emerge from mere aggregation of individual corporate actions, especially when most of the corporate assets are owned outside Canada. This conclusion is most evident in the clash between the current national need to conserve fossil fuel resources and the desire of the various corporations involved not to have a substantial surplus above allowed sales contracts as this represents unrealized short term profits. Thus even at a time when Canada is projecting serious oil and gas shortfalls from conventional low-cost sources by the mid-1980s and mid-1990s respectively, spokesmen for the oil and gas corporations are expressing concern about temporary surpluses possible in the interim and urging government to allow increased sales.[8] This is in the tradition of the 1960s when the National Energy Board used highly inflated potential reserve figures for oil and gas, supplied by the private corporations, to permit high volume sales to the United States and the continued encouragement of high energy-consumption policies in Canada.

The National Energy Board's reliance, at that time, on data supplied by the private corporations illustrates another reason for concentrating on public policy statements: the privacy of private corporations. Neither we, nor often, it seems, our governments, are aware of the decisions and information internally available to these corporations. This is perhaps appropriate to their status as private, competing individuals in the market, but it constitutes an important reason for insisting on government responsibility for energy policy if there is to be ultimate public control over so fundamental an aspect of life. Investigation of the interests and decisions of these corporations is a mammoth task extending well beyond the scope of this book, but the interested reader might well begin with the several studies published in recent years.[9]

At present a desirable level of public accountability by government, let alone by private corporations, is well nigh impossible because of the chaotic state of communication about energy matters. There is no national or provincial system of energy accounting. Some of the most crucial data have either never been systematically collected or are estimated by various institutions with conflicting results. For example, there are still only one or two national estimates of energy consumption given in terms of the temperature of the use involved and these are hidden away in obscure research papers. There are no official provincial estimates of this sort for Ontario at all, even for the relatively better known and publicly controlled electricity sector. To acquire the available data, incomplete and inconsistent though it be, an interested party currently has to consult more than thirty government research documents, many of them not ordinarily available to the public. Even when acquired, the data are aggregated in different ways – by energy source, resource ownership, primary demand, secondary demand, or economic sector demand – and they are not easily related to one another. In fact we consider it one of our important contributions to the quality of the public debate to set out some of these data in one place. This will at least allow others the opportunity to form their own conclusions on the available evidence. It is, however, a poignant commentary on the newness of the realization of the social importance of energy and energy policy that the relevant information has not yet been brought before the public.

We aim to stimulate others to form their own conclusions. We have not attempted to hide the complexity of the issues involved, but to present them as clearly as possible. Though the directions in which our preferences lie are clear, we have tried to lay out the costs, benefits, and the uncertainties sufficiently well to allow independent decision. Moreover, the question is not just one of technological decisions made under some purely empirical criteria, but rather is one of the fundamental evaluation of the choices about the kind and quality of life to be lived. Because of this, it is not appropriate to present categorical 'solutions'; rather, we aim to stimulate a process of public debate that will lead to choice of an energy policy and of a policy-making process.

PART II
A CRITIQUE OF CONVENTIONAL ENERGY POLICY

2
Energy dilemma and policy response in Canada

ENERGY DILEMMA

During the past century, the amount of energy that Canadians consume has steadily increased until, today, per capita consumption of energy in Canada is second in the world only to that in the United States (Figure 1). During this period Canada has also enjoyed a high rate of economic growth, based in part on ready access to cheap energy resources. Canadians have generally acted as though energy would always be available in the forms required, in essentially unlimited quantities, at relatively little expense, and with low social and environmental costs. Such implicit assumptions continue to underlie Canadian social and economic activities. They are evident in car-oriented cities, poorly designed buildings, and supermarket packaging. Today, however, our nation is confronted with a serious predicament in the energy sector that challenges all of these traditional assumptions about energy. This predicament strikes at the very basis of our energy-dependent economic and social system.

The world's industrialized nations developed an immense and diversified economic system over a relatively brief period of expansion that was dependent upon inexpensive and readily available fossil fuel energy. The evidence before us today indicates that that era is over. Much of the evidence was summarized in a 1976 federal energy study entitled *An Energy Strategy for Canada: Policies for Self-reliance*.[1] Briefly, the report stated that our demands for energy are continuing to rise rapidly at a time when existing, readily available and relatively inexpensive reserves of oil and gas are rapidly being depleted (Figures 2–5). New discoveries of oil and natural gas have been disappointingly small to date,[2] and new resources are increasingly sought in regions so remote and environments so hostile as to present immense tech-

FIGURE 1

A / Energy Consumption per Unit of Gross National Product (GNP)
(Source: Adapted from U.N. *Statistical Yearbook, 1975*)

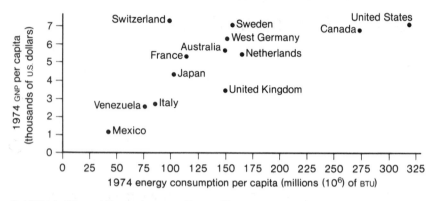

B / Effect of Import Dependence on Energy Use
(Source: Adapted from Royal Commission on Electric Power Planning, Interim Report,
A Race against Time [Toronto, 1978], p. 16)

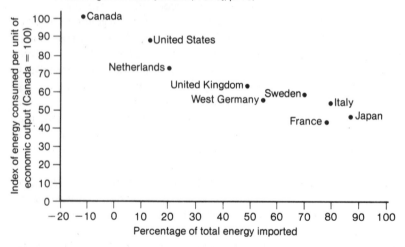

nical, economic, and socio-environmental difficulties in producing and transporting them to market (Table 1, p. 29). Most large and conveniently located hydro-electric power sites have been developed, and the remaining large sites are generally located in remote regions of the country (James Bay, Nelson River) where development presents similar difficulties. Capital costs

FIGURE 2

Demand for Energy, by source

(Source: Adapted from Energy, Mines and Resources Canada, *An Energy Strategy for Canada* [Ottawa, 1976], p. 52, Figure 4, by valuing primary electricity at its output value of 3,412 BTU per kilowatt-hour)

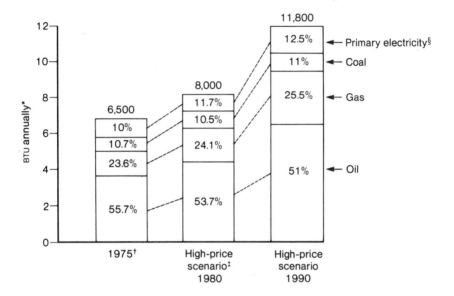

* Energy in 10^{15} BTU equivalent
† Estimated. 1975 and 1976 consumption figures and 1977 estimates differ by less than 2 per cent of these shares. See, for example, Gander J.E. and F.W. Belaire, *Energy Futures for Canadians* (Ottawa: Energy, Mines and Resources Canada, 1978), p. 303 (but valuing electricity at 3,412 BTU per kilowatt-hour).
‡ Assuming that Canadian domestic energy prices will rise to international levels by 1979, a substantial increase over 1975 prices. As of 1980 this had not yet occurred, reducing pressure to lower demand.
§ Primary electricity is electricity directly generated by hydro-electric and nuclear generating plants. In 1975 nuclear-generated electricity represented about 0.5 per cent of the 10 per cent total.

of energy production facilities such as electrical power stations, tar sands plants, or coal production facilities are rising steeply; obtaining the necessary capital is increasingly difficult and threatens to distort the economy (Tables 2,3, p. 30). Political issues about native land rights, provincial ownership of energy resources, mounting environmental disruption by exploitation of

FIGURE 3

Demand for energy, by use
(Source: Adapted from Energy, Mines and Resources Canada, *An Energy Strategy for Canada* [Ottawa, 1976], Figure 5, p. 53. Electricity is valued at its output value of 3,412 BTU per kilowatt-hour)

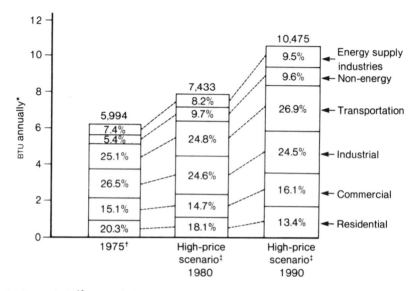

*   Energy in 10^{15} BTU equivalent.
†   Estimated. 1975 consumption and 1976 estimates given by Statistics Canada vary by less than 1½ per cent from these figures in all categories except commercial, where the variation is 3 per cent. Cf. note to Figure 8.
‡   Assuming that Canadian domestic energy prices will rise to international levels by 1979, a substantial increase over 1975 prices. As of 1980 this had not yet occurred, reducing pressure to lower demand.

energy resources, the safety and public control of such 'high-technology' energy systems as the nuclear-electric system, and international trade in energy resources and its potential military-political consequences are all pressing to the fore of the national consciousness.

While the growing shortfall between supply and demand could perhaps be met by importing energy, this option carries with it economic and political risks the government of Canada rightly views with great concern. The problems associated with the 1973 Organisation of Petroleum Exporting Countries (OPEC)[3] oil embargo and their recent pricing policies have vividly demon-

FIGURE 4

Domestic demand for and availability of oil*

(Source: Adapted from Energy, Mines and Resources Canada, *An Energy Strategy for Canada* [Ottawa, 1976], p. 83)

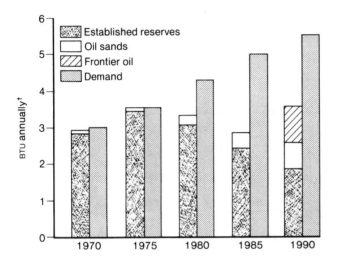

* Assuming that Canadian domestic energy prices will rise to international levels by 1979, a substantial increase over 1975 prices. As of 1980 this had not yet occurred, reducing pressure to lower demand.

† Energy in 10^{15} BTU equivalent

Note: Post-1976 federal energy publications clearly indicate that it is no longer part of federal government thinking to expect frontier oil to emerge before 1995 and even its still longer-term future is now open to serious question. See, for example, Figure 7.

strated the vulnerable position in which countries with a major dependence upon imported energy find themselves. On its present course, Canada faces an oil import bill running up to or above $9 billion per year by 1985 and a cumulative bill of around $90 billion or more by 1990.[4] OPEC price increases in 1979 undermined any hope that these projections might be exaggerated. In comparison, Canada's substantial gas exports to the United States are only expected to come to a third of this amount over the same period.[5] Moreover, there is real uncertainty about the future price of oil on the international market. Several studies predict the peaking and subsequent decline of world oil supply in the 1985–90 period when world demand for oil will be higher than ever and is expected to rise rapidly.[6] In these circumstances, oil prices could rise even more substantially, adding an even greater burden to

FIGURE 5

Domestic demand for and availability of natural gas*
(Source: Adapted from Energy, Mines and Resources Canada, *An Energy Strategy for Canada* [Ottawa, 1976], p. 84)

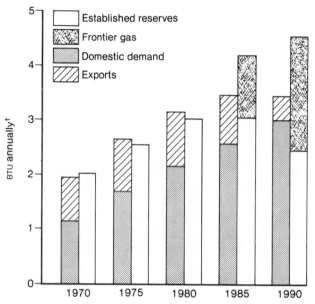

* Assuming that Canadian domestic oil prices will rise to international levels by 1979, a substantial increase over 1975 prices. As of 1980 this had not yet occurred, reducing pressure to lower demand.
† Energy in 10^{15} BTU equivalent.

Canada's trade deficit. In addition, the possibility that oil may be unavailable at any price should not be discounted entirely. The cutoff of Iranian oil and the difficulty Canada experienced in obtaining increased supplies from other established sources (e.g., Venezuela) should serve to underline the vulnerability of any energy policy based solely upon reserve estimates, however reliable, without regard to social and political developments.

It is difficult, of course, to evaluate even these uncertain import figures in the absence of other information such as the inflation rate and the path of industrial development. However, Canada's economic performance over the past decade does not suggest a bright future for its economy.[7] In the context of a static or deteriorating export performance, it seems reasonable to

TABLE 1
Capital investment requirements to bring energy to market,
various sources, 1975 (in billions of 1975 dollars)

Energy source	Investment / 10^{15} BTU
Refinery for offshore crude	0.6
New Prairie oil	
Production	1.12
Refining	0.60
Total	1.72
Frontier oil	
Production	2.16
Transport	1.55
Refining	0.60
Total	4.31
Alberta oil sands	
Production	2.84
Refining	0.6
Total	3.4
Frontier gas	
Production	0.41
Transport	1.31
Total	1.72
Gas from coal	4.31
Nuclear station	6.04

Note: Although optimistic, these figures show the ten-fold
increase in capital investment required in moving from
conventional oil to nuclear energy. These figures might
be compared with the somewhat more pessimistic estimates in A. Lovins, *Soft Energy Paths* (Cambridge,
Mass.: Friends of the Earth/Ballinger, 1977), p. 134.

Source: Adapted from R.M. Dillon, *Ontario: Towards a Nuclear
Electric Society?* (Toronto: Ontario, Ministry of Energy,
October 1975).

conclude that foreign exchange deficits will constitute a serious economic
problem.

In all, the energy future Canadians face promises serious economic, technical, social, and environmental problems, aggravated by the magnitude of
present energy consumption and its expected growth. These problems are
combining to produce a situation in which our best efforts to supply energy

TABLE 2
Estimated energy-related capital requirements,* various sources, 1976–90
(in billions of 1975 dollars)

Energy source	1976–80	1981–85	1986–90	Total
Electric power	21.7	32.5	37.0	91.2
Pipelines	5.7	19.2	3.0	27.9
Petroleum				
Exploration and development	10.6	15.5	14.2	40.3
Refining and marketing	3.9	4.1	4.8	12.8
Oil sands	3.0	2.3	0.3	5.6
Coal	0.8	1.5	0.9	3.2
Energy investment	45.7	75.1	60.2	181.0
Estimated GNP	912.1	1,162.0	1,438.5	3,512.6
Energy investment as percentage of GNP	5.0	6.5	4.2	5.2

* These figures are based on the assumption that Canadian domestic energy prices will rise to international levels by 1979, a substantial increase over 1975 prices. As of 1980 this had not yet occurred, reducing pressure to lower demand.

Source: Adapted from Energy, Mines and Resources Canada, *An Energy Strategy for Canada* (Ottawa, 1976), p. 108.

TABLE 3
Research and development expenditures on energy, various sources
(in millions of current dollars)

	1975–76	1976–77	1977–78	1978–79	1979–80
Conservation	8.6	8.0	11.0	12.5	12.5
Oil and gas	10.5	12.3	8.5	11.4	12.6
Coal	2.8				
Nuclear*	84.8	90.3	87.9	105.8	106.4
Renewable†	1.6	4.6	5.4	14.5	19.4
Transport and transmission	4.9	5.2	4.4	6.1	6.1
Miscellaneous‡	–	.2	1.0	.5	.8
Total	113.2	120.6	118.2	150.7	157.9

* Includes uranium resources.
† Hydro and tidal, solar, wind, geothermal, and biomass.
‡ Includes co-ordinations.

Source: Adapted from Energy, Mines and Resources Canada, *An Energy Strategy for Canada* (Ottawa, 1976), p. 145 and 'Energy Conservation in Canada: Programs and Perspectives' (EP 77–1), p. 47, and private communications with EMR.

FIGURE 6

Domestic demand for and availability of total energy*
(Source: Adapted from Energy, Mines and Resources Canada, *An Energy Strategy for Canada* [Ottawa, 1976], p. 82 by valuing electricity at its output value of 3,412 BTU per kilowatt-hour)

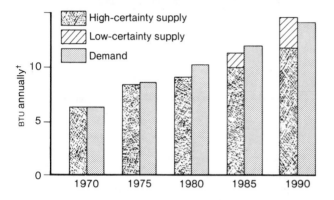

* Assuming that Canadian domestic energy prices will rise to international levels by 1979, a substantial increase over 1975 prices. As of 1980 this had not occurred, reducing pressure to lower demand.
† Energy in 10^{15} BTU equivalent

Note: According to recent public discussions, even under very optimistic assumptions about the availability of 'Low-certainty' energy, Canadian aggregate self-sufficiency is only attained around 1990. However, Canada will remain in a deficit position with regard to oil, currently the most important form of energy, indefinitely into the future. Figures 4 to 6 leave unanswered what happens after 1990. Cf. Figure 7.

in the near and longer term future are unlikely to satisfy projected demands (Figures 6,7), and attempts to import the shortfall will only create further difficulties of similar sorts. Though we take issue with the importance the recent comprehensive analysis of future energy requirements by Energy, Mines and Resources Canada attaches to electricity as an appropriate panacea to these difficulties, we concur with the two leading conclusions of the report: namely, that the next twenty-five to fifty years will see a developing crisis in the provision of fossil fuels, and that the major limitation to effective responses to this development is the constraint on the rate at which various energy forms can be substituted for one another in the economy.[8] We shall take pains to stress these same points.

FIGURE 7

The supply gap for total energy
(Source: Derived from C.M. MacNabb, 'The Canadian Energy Situation in 1990," Third
          Canadian National Energy Forum, Halifax, 1977, Figure 16; however, we have
          reduced the electrical supply component from 10^4 to 3,412 BTU per kilowatt-hour
          and reduced the demand and demand-conservation curves by the same amount
          as was thereby subtracted from the supply total)

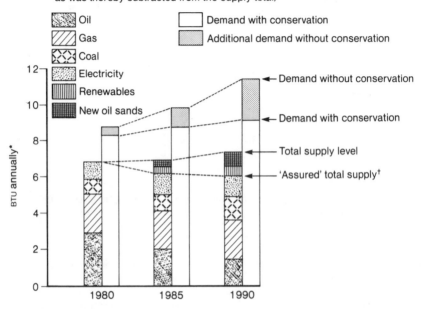

* Energy in 10^{15} BTU equivalent.
† 'Assured' supplies refer to energy available from proven reserves and existing and committed
   production facilities of all types but do not include any additional projects. The demand-
   conservation curve assumes a vigorous, *enforced* federal energy conservation programme in
   effect (a new element in federal policy). Cf. Figure 9, curve c, and text. By contrast with
   Figure 6, here energy self-sufficiency for Canada is pushed off from 1986 at least until 1996.
   Moreover, depending upon assumptions about 'non-assured' supplies and conservation, a
   balanced national energy budget may be postponed indefinitely into the future.

Thus major changes in our pattern of energy use and in related social
policy are required to cope with the situation now emerging. Since every
human activity is powered by an energy flow, we can expect that changes in
the pattern of energy use will require corresponding changes in almost every
important area of life. For example, the decline of oil and gas availability
spells rapidly increasing costs for motor vehicle and air transportation, for oil

and gas home heating, and for food because of rising costs for fertilizer, fuel, and food processing.

More specifically, plans for massive electrical development in Ontario will affect our ability to finance other economic development because of the drain on capital markets. They will also very likely nurture the already highly centralized, large, and impersonal institution that is Ontario Hydro, whose decisions will have a strong influence on the pattern of urban and regional development, employment opportunities, and future energy options.

PATTERNS OF ENERGY USE

To understand the nature of the energy dilemma facing Canada today, it is important to be familiar with the historical patterns of energy production as well as with current patterns of energy utilization. The last hundred years have seen several major shifts among energy forms. Starting from its initial and long-time dependence upon renewable forms of energy – wood and waterpower – Canada has steadily developed a much greater dependence upon non-renewable energy forms, primarily oil, natural gas, and coal. In 1979 these fossil fuels provided 90 percent of all energy needs. The most important physical factor in the Canadian energy predicament is the decline in oil and gas supplies and the difficulties associated with substituting either coal or electricity for them.

Further, very rapid and accelerating growth in energy consumption has taken place during the last fifty to seventy-five years. In particular, in the period before 1972, Canadian primary energy demand was increasing at about 5.3 percent annually.[9] This rate of growth would, if continued, result in a quadrupling of the 1972 energy demand before the year 2000. After 1972, rapidly changing economic circumstances, including the Middle East oil embargo, price increases, and uncertain government policy, resulted in a temporarily reduced rate of growth in energy consumption along with an overall depressed economy. Given the existing economic and technological framework, we may reasonably expect that a general pickup in economic growth will lead once again to rapidly rising energy demands. Even so, in the 1972–77 period, total primary energy demand grew by 22 percent and electrical demand by about 27 percent. Rapid growth in energy demand makes it so much harder to respond adequately to the decline in fossil fuels; even without their decline it would be almost impossible, and certainly socially disruptive, to continue to meet an annual growth in demand for energy of 5.5 percent.

In the past, too, each form of energy, in effect, provided the energy basis for the process of transition from its own use to the succeeding new energy form. Wood continued to be used during the time required for coal-burning technology to be widely adopted. Likewise, coal-based energy technology supported the economy while oil-based technology was developed and applied. Thus it is imperative that Canadians act sufficiently far in advance of the decline in major fossil fuels so that there will be sufficient fossil energy to allow for the transition period from the current to an alternative pattern of energy use with a longer future.

Experience suggests that approximately thirty to fifty years are required to achieve the transition from a major dependence upon one form of energy to a major dependence upon a different energy form.[10] This is a long lead time, and it indicates something of the fundamental political challenge that energy policy poses. If short-term economic and political expediency is pursued, then long-term difficulties will mount. Since choices among energy policies involve basic social and ethical decisions, democratic processes will be challenged as never before.

Finally we note that energy problems are aggravated by the inefficient use of energy. As we pointed out, in comparison with other countries only the United States consumes more energy on a per capita basis. In 1975 Canadians used the energy equivalent of about 3.75 million barrels of oil per day. This works out to about 60 barrels (2,100 imperial gallons) of oil per person each year. Many industrialized countries throughout the world use significantly less energy per capita. In particular, Sweden consumes less than two-thirds the energy per capita to achieve essentially the same gross national product per capita as Canada, though both countries have similar economies, climates, and styles of living.[11] Looked at in a different way, this high degree of Canadian inefficiency of energy use offers significant opportunities for reducing energy demand merely by adopting measures already widely used in countries with a less fortunate history of access to cheap energy resources.

POLICY RESPONSE

The major emphasis of federal policy during the next fifteen years is to be concentrated upon strenuous attempts to increase the supply of conventional energy, principally oil, natural gas, and electricity, including a significant fraction to be generated by nuclear energy (see Tables 2, 3, and 4).[12] Such a policy will be costly, even for a country as wealthy as Canada because of rising capital costs (see Table 1). Whereas in the past about 3.5 percent of gross national product (GNP) was directed into making energy available, the capital required during the next fifteen years may rise to as much as 6 percent of GNP representing an expenditure of $180 to $210 billion. This figure is

TABLE 4
Projection of installed electrical capacity, 1975–90 (in megawatts)

	Hydro	Nuclear	Other†	Total
1975*	36,800	2,660	20,080	59,540
1980	42,500	6,200	23,700	72,400
1985	55,500	12,800	29,500	97,800
1990	66,200	22,200	36,800	125,200

* Estimate.
† Other generation includes electrical power generated from coal, oil, and natural gas.
Note: The low-growth case presented here was generated within Energy, Mines and Resources
    Canada and represents a capacity expansion programme to meet load growths that
    increase on average by 5.5 per cent per year from 1976 to 1990. A high-growth case is
    considered, based on provincial electric-utility projections as at end-1975, that would add
    roughly 20 per cent to the 1990 total.
Source: Adapted from Energy, Mines and Resources Canada, *An Energy Strategy for Canada*
    (Ottawa, 1976), p. 68.

from $60 to $90 billion (in 1975 dollars) in excess of the average historical
rate of expenditure on energy supply. The excess alone represents an annual
expenditure on energy of something like 50 percent of the total 1975 expenditure on residential construction.[13]

Faced with such a vast programme, a critical question that must be
resolved is whether or not the proposed policy direction will result in a more
secure and satisfactory energy situation for Canada in the years ahead. In
short, does the proposed policy make long-term sense?

While *An Energy Strategy for Canada* outlines the consequences of the
proposed programme in terms of supply and demand of energy up to 1990,
the post-1990 implications of pre-1990 policies are not discussed. It is possible, however, to analyse a number of implications that the proposed policy
has for our energy future and to criticize it on the basis of several criteria. Will
the policy succeed on its own terms, economically, and in the long term?

Unfortunately, the proposed strategy will be unable to prevent indigenous
energy supplies from falling further and further behind the projected demand
for energy (see Figures 4 to 7). Only under utopian circumstances such as
appropriate technological breakthroughs, success in exploration, and the
ready availability of capital, manpower, and equipment does the policy envisage closing the demand/supply gap.[14] In short, the proposed policy, which
still anticipates supply shortfalls at least until the mid-1990s and perhaps
beyond, rests on numerous uncertainties.

Consider first the future of oil supplies. A mounting oil supply deficit from Canadian sources and a corresponding rise in oil imports and capital outflows in payment are anticipated. To decrease this shortfall, or at least restrain it from further increase, official policy relies first on new oil discoveries resulting from increased exploration. These yields are, of course, uncertain and all the associated costs are rising rapidly. Discoveries to date have been disappointingly small, and increasingly federal policy has come to rely on exploiting oil sands and heavy oils, principally from the Athabasca tar sands deposits of western Canada.[15] In fact, rapidly rising yields from this source seem at best a questionable proposition. Although reserves of oil are known to exist within the tar sands, there is uncertainty about when, and even if, any additional extraction plants will be constructed following the present Syncrude plant. Some of the factors contributing to this uncertainty are the very high capital costs involved in constructing oil extraction plants (now in excess of $3 billion), the management difficulties involved in constructing such massive plants within the time span involved, and some remaining doubts as to the practicality of the still unproved extraction technology. Thus, not only is the supply of imported oil unpredictable, but also the indigenous supply of oil is equally uncertain.

The exploitation of tar sands oil offers an excellent example of the kind of conventional supply-oriented policy that dominates current energy policy. Mr G. MacNabb, deputy minister of Energy, Mines and Resources in 1977, summed up the prospect in these terms:

And now let's really be bullish and assume we can overcome the problems preventing a rapid development of oil sands and heavy oils. We pay the price required; we develop new fiscal systems; we build upgrading plants; we overcome environmental and social problems; and we resolve federal-provincial differences. What do we have? Conceivably one million barrels a day by 1990. Capital cost? – in the range of 15 billion dollars in today's dollars. Operating labour force? – up to 20 thousand. Obviously this is an optimistic target, but when we enter it into the equation, it does little more than compensate for the drop off in production capacity from our established reserves.[16]

The expenditure of $15 billion in twelve years indicates just how visionary the 'new fiscal systems' might have to be. When we understand that about eight of the existing Syncrude plants would be involved and they would accumulate tens of thousands of tons of sulphur annually, in addition to much greater quantities of oil-contaminated waste water and tailings from the extraction process and other wastes, we gain some impression of the likely environmental problems. The nation-wide, regional, and local impacts of such a huge influx of men, money, and materials into one small area are likely to be significant. The federal-provincial concessions and transfer of power necessary to undertake such an effort, in the light of current federal-provincial conferences about control of resources, should give less energy-favoured regions cause for concern. All this activity will only contribute an increase of less than one-quarter percent to national employment and a rise in energy supply that barely compensates for the predicted falloff in oil production. Does this kind of investment constitute a rational energy policy? Are there not more effective energy investments? Possible solutions to some of these questions may be found in the discussion of alternative energy policies in Part III.

The situation with regard to natural gas is portrayed as being somewhat brighter, but the uncertainties remain. All that can be said is that *if* the frontier reserves of natural gas that are hoped to exist can be located and subsequently delivered to market, Canada may have a supply of natural gas in excess of its national requirements in the late 1980s (e.g., as indicated by the 'frontier bubble' in Figure 5). But any production of natural gas in the far north and offshore is likely to be both expensive and potentially damaging to the environment. In addition, a number of issues – native land rights and future domestic control of gas exports – remain unresolved. In short, economic, resource, environmental, technical, and social constraints exist that lead to future uncertainties about the availability and quantity of frontier gas.

Recent natural gas discoveries in Alberta, rather than the Arctic, have gone some distance towards encouraging optimism. But they should be considered cautiously for while Energy, Mines and Resources evidently felt confident in 1978 that supply would exceed total demand through 1990, the National Energy Board projected a shortfall as early as 1984.[17] Nor is there any guarantee that the economic conditions would be appropriate to allow substitution of gas for oil.[18] Moreover, and crucially, such 'bubbles' create their own economic pressure to sell the gas and in fact to export rather than conserve it. In April 1978 Canada agreed to permit the export of more than 350 billion cubic feet of gas to the United States annually for the next ten years and in December 1979 the federal government agreed to export almost twice that amount (3.75 trillion cubic feet) to the United States over the 1980–87 period. And still there is no guarantee of eventual access to Alaskan gas, although the Canadian public had been led to believe this would be a precondition of increasing exports. It is interesting to speculate about the politics involved.[19]

Though Canada possesses large reserves of coal, it has been phasing out coal-burning technology for over thirty years. With the exception of a few industrial applications and a minor role in the generation of electricity, coal is no longer used. To attempt to return to the use of coal in residential space heating, rail transportation, and commercial steam generation would be expensive and time-consuming. Moreover, there are serious environmental consequences from both strip and underground mining and problems of air pollution and ash disposal after burning. Newer, cleaner coal-burning technologies exist or are now in development but these will substantially increase the cost of the resulting energy.[20] While the production of coal in Canada could be increased significantly without major technological innovation, the corresponding expansion of its use to replace other fossil fuels is problematical, especially in the short term, if only because of 'acid rain'.

A major focus of the federal policy is the construction of large amounts of hydro-electric, fossil fuel, and nuclear electric generating capacity (see Table 4). This construction is to absorb at least 50 percent or $91 billion of the planned investment in energy supply technology over the next fifteen years. Such installations are highly capital intensive, and even after this expenditure the total electrical contribution to the 1990 national energy budget would be only 12.5 percent, a rise of only 2.5 percent over the 1975 contribution.[21] Even this rate of construction will impose severe demands on capital, materials, and skilled manpower. Thus the data indicate it will be very difficult to build electrical generating plants rapidly enough to compensate for the energy deficits incurred by oil shortfalls.

From this perspective, it is important to understand that not even nuclear energy provides a solution to the problem of inadequate energy supplies. Not only is the rate of nuclear expansion limited, but its present uranium base, technology, and economics are all subject to important uncertainties (see Chapter 5).[22] In addition, oil is used in a number of applications, such as transportation, lubrication, and petrochemicals, in which it would be technically extremely difficult for electricity, however generated, to play a major role in the near future. Thus electricity cannot be viewed as a panacea for energy supply problems and even its proposed rate of growth is subject to a number of significant uncertainties.

Against this rather pessimistic backdrop, the recently increased federal interest in a conservation component to the policy package is understandable. Federal energy policy in 1978 assumed a lowering of annual growth in primary energy demand to 2 percent through vigorous conservation measures (see Figure 7). In contrast to this stands the earlier federal conservation target to reduce annual growth to 3.5 percent (Figure 6); the 2 percent target was only discussed then as a more distant possibility. However, both conservation targets are modest, passing little or not at all beyond what the inevitable price rises and slow growth of the 1980s may bring.

The effects of price rises on demand are not to be underestimated. They have played a major role in previous shifts in energy sources and technologies and will remain a key factor. However, pricing is now heavily influenced by social policy decisions, aimed at goals other than short-term competitive market equilibrium, and for good reason. In the energy planning field, the range of potential social action extends well beyond what narrowly determined market prices might permit at any given time.

Though not so poorly treated as renewable energy resources, conservation is not a major government priority (see Tables 2 and 3), relative to the need to reduce energy demand. And most important, the addition of the contemplated conservation component will not solve the energy supply problem, as it is not sufficient to achieve energy self-sufficiency for Canada (see Figure 7). Fortunately, detailed studies suggest that these conservation measures fall short of what is reasonably possible (see Chapter 8).

The subsequent 1978 federal study *Energy Futures for Canadians: Long-term Energy Assessment Program* (LEAP) virtually repeats the supply orientation of the earlier energy strategy paper, with the difference that there is a substantial shift of emphasis from oil to nuclear electricity and that renewable energy resources are given a minor, though noticeable, role to play. Up to the year 2000 conservation targets are not even as stringent as the 1978 call for a 2 per cent annual growth ceiling (LEAP hopes for about 2.7 per cent); from 2000 to 2025 the conservation target is around 1.3 per cent, still well below what is reasonably achievable for that period. All told, the main difference LEAP adds to the federal policy picture is a 'business-as-usual' perspective that stretches thirty-five years further into the future than did the strategy document. We can see no evidence that it escapes any of the difficulties already laid out, or yet to be laid out, for such a policy approach.[23]

Our conclusion is that no mix of conventional energy resources can realistically satisfy projected energy demands, even to 1990, and that the future of each is subject to important uncertainties. Taken together the constraints on, and uncertainties about, each energy resource, coupled with rising energy demand, make it unlikely that the present federal energy policy will lead to an energy self-reliant Canada.

The economics of present policy
It is also important to understand the broader economic implications of the federal strategy. Although the topic is complex, enough can be outlined in a fairly simple manner to justify a pessimistic judgment.

In contrast stands the apparent optimism of government. The federal government evidently does not anticipate that the large capital expenditures envisaged by its energy policy will disrupt the smooth operation of the national economy. Though projected economic and demographic growth rates show a decline from the trends that prevailed in the 1960s, the government energy policy assumes that gross national product and industrial output will more than double by 1990 with personal disposable income growing by almost 50 per cent.[24] Notwithstanding the fact that the economic projections were obtained by the use of sophisticated econometric models, a careful reading of the arguments and assumptions suggest several reasons for seriously questioning their viability.

To begin, the energy sector will require very large sums of money from foreign sources; yet there is a reluctance to face up to the problems such a strategy may create. When asked at a public meeting what proportion of the forecast $180 billion of capital investment will be supplied from foreign sources the deputy minister of Energy, Mines and Resources replied: 'You tell me, I have no idea. It all depends on the willingness of Canadians to invest in high risk energy developments.'[25] Moreover, the figure of $180 billion should not be regarded as certain. For example, R.A. Utting, general manager of the Royal Bank of Canada, put the figure at $300 billion for the period to 1990.[26] Beyond that time the LEAP document concedes that the annual rate of investment will need to rise even higher for the 1990–2000 period and after, though no definite figures are provided.[27]

If these funds come as direct foreign investment then this will increase the already excessive degree of foreign ownership of the Canadian economy. In this regard we note the growing trend of foreign-owned Canadian-based multinational energy corporations to reinvest their rapidly growing Canadian profits in acquiring ownership of non-oil energy forms such as coal and uranium in Canada. Substantially increased foreign loans will increase the borrowing institutions' susceptibility to foreign influence. It is also likely to limit their future borrowing capacity and increase the risk of default on repayment. In any case, there may well be a shortage of long-term funds for capital-intensive energy technology, nationally and internationally, since Canada is not the only 'wealthy' country facing such huge expenditures for energy.[28] Also, it is generally agreed that short-term debt is an inappropriate fiscal instrument for funding long-term investments, because it is expensive and unreliable.

Second, Canada's deteriorating performance in current trade in manufactured goods and services will only aggravate the need for foreign capital. In the past decade, Canada has become increasingly dependent on the import of manufactured goods and services for which it has partially paid by exports of raw materials. The Science Council of Canada has expressed concern about 'the weakness of Canada's manufacturing sector' and notes that 'the country, industrially, is rapidly falling behind other nations, and by default, placing its hopes for the future on a resource sector, which, in its present form, is inadequate to the task of raising or even maintaining the standard of living which most Canadians take for granted.'[29] Although the Science Council study does see some 'patches of light,' it would seem unlikely that Canada will be able to reverse a trend dating back at least to 1955 in sufficient time to provide a trade surplus with which to fund the necessary investments in energy. Rather, it is to be expected that Canada will continue to rely on capital imports to finance its deficits on current account, and these capital imports will be magnified by the added requirements of the costly investments in energy supply.

In its own assessment of the problems involved in financing energy development, Energy, Mines and Resources foresees a reduction in government borrowing because of increases in surpluses from the Canada and Quebec pension plans and a decrease in unemployment insurance payments. The department used optimistic assumptions about continued economic growth and a marked reduction in unemployment and inflation.[30] Commenting on this scenario, Stephen Duncan of the *Financial Post* wrote: 'If the world unfolds in this particularly pleasant fashion, then Canada, in the late 1980's, becomes a net lender on foreign markets, rather than a net borrower. Pollyanna would approve.'[31]

Another detailed analysis of the availability of capital to fund the development of Canadian energy supplies states that the necessary funds could be accumulated 'partly by an increase in the aggregate scale of investment, but even more by a shift in the pattern of domestic investment relative to current dollars GNP.'[32] Summarizing this work, Duncan remarked that 'analyses of this kind may convey the unintended impression that the projected adjustments of investment levels and patterns will be fairly smooth and readily accomplished. Nothing could be further from the truth.'[33]

Third, despite the stated policy of the federal government to keep down the share of the government sector in the gross national product, the federal energy strategy projects a substantial increase in the proportion of GNP taken in taxes. It is unlikely, therefore, that the government would want to add further increases in taxes to provide funds to the energy sector.

Fourth, shortages in skilled labour and materials may prevent successful completion of the energy investment programme and/or significantly raise its cost above current estimates. For example, there are tens of thousands of skilled jobs presupposed in even the modest schedule for Alberta tar sands production projected in Figure 7. In January 1978, then acting Deputy Minister of Energy, Mines and Resources, G. MacNabb, admitted that finding a qualified labour force was one of the severest constraints on the enterprise. Similar problems likely will confront the nuclear-electric industry.

In sum, the future sketched out in *An Energy Strategy for Canada* does not offer a solution to Canada's energy problem. Pursuit of the strategy will place increased pressures on an already fragile economy.

Long-term implications
The 1976 energy strategy takes no account of the implications of its proposed policies in the post-1990 period. Clearly, these policies lead to a continuing and a growing dependence upon oil and natural gas, with the result that in 1990 Canada would still be overwhelmingly dependent upon fossil fuels, especially oil and natural gas. Moreover, Canada will be consuming these fuels much more rapidly than it is today. On the basis of the federal projections of energy demand, Canada would be consuming 2.39 million barrels of oil a day in 1990, 55 per cent more than in 1975. Consumption of natural gas will have increased by 98 per cent over the same period.

Such a policy of continued and growing dependence upon fossil fuels is alarming in view of the wide recognition that, within approximately thirty years, the era in which oil and gas dominated will likely be over. The fact will apply both domestically and internationally and with a vengeance if energy planning does not compensate.

Historically, industrialized economies have typically required thirty to fifty years to make a transition from a major dependence upon one form of energy to another. Thus, any policy that perpetuates and increases dependence upon fossil fuels would appear to make sense only if the nation is confident today that in 1990 there will exist – and be available – the very large additional reserves of oil and natural gas required to support this country for the 30-year transition to other energy forms. Although the 1976 federal study admits the need to begin the lengthy transition from natural gas and oil immediately, it shows no serious recognition of the importance of using our remaining oil and gas resources with great care as the energy basis for this transition. Indeed, it does not discuss the necessity of transition reserves of oil and natural gas at all.

The other major component of federal supply policy is electrical energy, which includes a large nuclear component. This technology also has its uncertainties and disadvantages, some of which have already been mentioned. One of its chief long-term disadvantages is its inflexibility. Nuclear-electric plants can only use nuclear fuel, and the eventual decline of economically attractive uranium supplies is as assured as is that of oil. This decline is perhaps not more than thirty years further into the future than that of oil. At that point the investment in nuclear technology will be vast indeed. It is likely then that new and as yet untried nuclear technologies will be required that will add a new round of investment expenditures (see Chapter 6). Moreover, plausible alternative nuclear technologies, such as thorium/plutonium fueled or breeder reactors are likely to be associated with still larger social and environmental risks. The LEAP report only shifts the balance between dependence on fossil fuels and nuclear electricity; it does not challenge their dominance. We conclude that any policy that offers more options would surely be preferable. What the federal policy offers us is a further entrenchment of our historic dependencies upon non-renewable energy resources at very large and mounting economic and social costs, with no allowance for transition from present resources to others or for their eventual depletion, and relatively little effort allocated to managing energy demand.

Environmental and social effects
The federal government expects that any substantial additions to our oil and natural gas reserves in the coming decades will come from the frontier areas: the Arctic and, to a lesser extent, non-Arctic offshore deposits. But exploration and drilling for gas and oil have already had a significant impact on the social and natural environment of the Arctic, even though exploration activity is well below the levels to be expected a decade hence. Thus the future implications are for further detrimental effects on the environment and existing society.

Exploration, the transport of oil and gas through pipelines, and the maintenance of equipment require the opening of large portions of Canada's north. This will unavoidably lead to the replacement of the traditional ways of northern life by a markedly different culture. Native peoples, hindered from supporting themselves in their accustomed ways, will be forced together in reserves, as were their southern relatives. Understandably there is a growing resistance to such schemes, and the wish of the native peoples to control their own destiny may become the major impediment to northern 'development.'[34]

The natural environments of the tundra and the northern inland and coastal waters will all suffer considerably from oil and gas exploitation, for their natural ecosystems have evolved under virtually identical, and extreme, conditions for long time periods, and they are exceedingly fragile. One of the consequences of present policy will almost certainly be the disruption of large tracts of a unique environment. This may have unexpected effects on the climate, the hydrology, and the wildlife of other parts of Canada, for very little is known about the exact role of the Arctic with respect to these and a large number of other factors. There is also the risk of major accidents. Nobody knows for sure what the result of a major oil spill in the Arctic would be. Indeed, it has been suggested that major climatic changes might result.[35]

Hydro-electric projects are only planned for a few, relatively isolated regions. But again, their scale and the social and environmental disruption that such projects cause have made them very controversial undertakings from any but a strictly technical point of view. The Nelson River and James Bay projects in Manitoba and Quebec respectively provide excellent examples of what happens. In addition, the increasingly 'desperate' hydraulic schemes that have been seriously discussed – most of which also involve sale of water to the United States – involve the reversal of northern Ontario rivers and the notorious NAWAPA scheme.[36] There might well be increased pressure to develop these to help overcome the post-1990 energy difficulties, and their potential for social and environmental disruption may be still greater than those already under construction.

It is not yet necessary to go to frontier areas for coal, and thus the threat to the environment may be thought to be less severe than that of development of oil and gas deposits. Yet large-scale open-pit mining is also notoriously destructive of the natural environment. And burning large quantities of coal without anti-pollution equipment produces air pollution, including 'acid rain', with all its consequences for human health and for the natural environment. Effective ways of controlling this air pollution already exist or are under development; their implementation, however, would likely require expenditures that would more or less cancel the present economic advantage of coal.[37]

In addition to specific locational effects, the relentless increase in environmental and social pressures in urban areas as the magnitude of the energy flows and the scale of the technology continually increases must be examined. Consider, for example, the effect of oil and gas transmission pipelines, the very high voltage electrical transmission lines now criss-crossing Ontario, the refineries and nuclear generating plants, and the continuous hauling and dumping of wastes, many of which are dangerous. Under federal policy these features can all be expected to intensify greatly, even to double by 1990.

By the turn of the century, energy technology will surround us all. Most of us will travel under, or over, its transmission systems daily, we shall live near one of its huge plants, and in most cases even take our rural recreation near them. In these circumstances, failure and the threat of failure take on a new dimension. We shall be surrounded by small failures, such as hydrogen sulphide leaks from nuclear heavy water plants and oil leaks from refineries and pipelines, and live more constantly under the threat of large-scale failure. Ironically, the chances of such accidents actually occurring cannot reliably be determined before we have experienced a good number of them. Whatever the ultimate impact on physical health, there seems little doubt that the psychological quality of life will deteriorate in subtle but important ways.

Equally subtle and no less important is the deterioration in the social environment we foresee. The threat of costly failure acts as a spur to enlarge police forces and grant them broader powers, all in the interest of preventing failure and responding effectively to emergencies. As centralization intensifies, so too does the rationale for a 'police state.' A bill was introduced to the Manitoba legislature, for example, in July 1980 calling for an energy police force with powers of search and seizure to enforce energy rationing in the event of a shortage or interruption in supplies. Violators would face fines up to $50,000 or up to two years in jail.

Moreover, massive, costly, high-risk energy production technology inevitably requires highly centralized, expert-dominated social institutions that increasingly impose a uniform way of life on the country and prohibit effective public control over energy decision-making. We can illustrate this briefly by considering the burgeoning electrical production system.

Management of the vast scale of present and planned electrical systems in a regionally integrated, highly centralized system requires considerable expertise that cannot usually be organized locally because the scale and design of the system require centralized management. These expensive and potentially dangerous installations represent increasing security risks requiring increasingly strict policing. Moreover, investments involved in planning, building, and operating such a system ensure that only a large utility with the financial backing of a provincial or federal government can effectively engage in electric power planning. The consequence is the removal of energy planning from regional and local control, which will lead to increasing regional dependence on central government and to the disappearance of the local idiosyncrasy and creativity associated with local autonomy. The ability to manipulate the future availability of electric power shifts political power towards higher levels of government.

Finally, we observe that the concentration of reserves of fossil fuels and uranium is highly unequal among Canada's provinces, creating vast inequities in wealth and energy security. These inequities have grown to the point where they constitute a major strain on the political unity of the country. By contrast, we observe that solar and forest biomass energy is distributed far more equitably; in fact, on a per capita basis, it tends to favour the less developed regions of the country. Increased reliance in the future on fossil and nuclear energy, then, can only lead in the first instance to increasing economic and political inequities, with all their ramifications.

In summary, even under the 'optimistic' federal scenario Canadians may look forward to increased dependence of their regions on federal and provincial governments and generally to an increased dependence on those institutions that control the planning and the functioning of energy supply systems. They may also expect to witness in most of Canada's north the forced disappearance of native peoples' traditional ways of life and the destruction of large parts of a unique natural environment. They will have to live in a much more intensely 'industrialized' environment and to live with increased risk of accident and the increased policing to which such threats give rise. Finally, Canadians can expect to live with the strains arising from greatly increased provincial inequities.

BUSINESS AS USUAL

Our analysis of available documents leads us to conclude that the federal energy strategy should be seen as a 'business-as-usual' policy because it leaves Canada with an even stronger dependence upon traditional non-renewable energy resources with no allowance for the longer term future; leads even more heavily towards the established dependence on foreign capital and towards ever more massive technical projects; reinforces the past trend towards less and less effective public participation in decision-making and towards a decreased sense of initiative and responsibility for social conditions; and leads towards maintenance of the present institutional arrangements with regard to energy policy and supply.

By contrast the long-term public interest calls for a sustainable resource base utilized as far as possible in the pursuit of social equality, diversity, and freedom of choice; social institutions that nurture responsible political participation and as much local autonomy as possible; and a flourishing, self-reliant economic life. Present policy supports the short-term interests of a small wealthy segment of society – the present energy technology industries – at the expense of longer term social goals.

3
Energy dilemma and policy response in Ontario

The array of energy-related problems facing Canada as a whole has especially serious implications for Ontario. That province consumes a larger amount of energy than any other, and fully one-third of the national total. Yet nearly all the energy marketed in Ontario is imported from regions beyond its borders, mainly from western Canada and the United States, in the form of oil, natural gas, and coal. In terms of these conventional forms of energy, Ontario is critically energy deficient – a 'have-not' province. The grave problems that confront many energy-poor nations of the world are facing Ontario today, as the province finds itself in a vulnerable situation in which control over the conventionally exploited forms of energy rests largely outside its jurisdiction.[1]

It may be argued that Ontario's energy deficiency is not a cause for concern. After all the province is situated within a national confederation of provinces and, doubts about the future of confederation aside, many of these provinces have more energy resources than Ontario. Increasingly, however, these provinces are implementing policies designed to put their own energy future before the needs of other provinces. Quotas, export restrictions, and increased royalties are ever more evident in interprovincial energy commerce.

Two important facts emerge from a survey of the basic data on the patterns of energy consumption in Ontario (see Figure 8 and Tables 5 and 6). First, by far the largest portion – or about 50 per cent – of the total provincial primary energy demand is met by oil. Oil, natural gas, and coal, almost all of which are imported from outside the province, provide nearly 90 per cent of all provincial energy needs.[2] Second, indigenous forms of energy that are now harnessed on a significant scale – hydro-electric power and nuclear-generated electricity – only provide about 9 per cent of total secondary

FIGURE 8
Approximate energy flows in Ontario, 1973, showing how the primary sources
(inputs) flow to the main sectors (outputs). Source, *Our Energy Options*, Royal
Commission on Electric Power Planning (Toronto, 1978), p. 16.

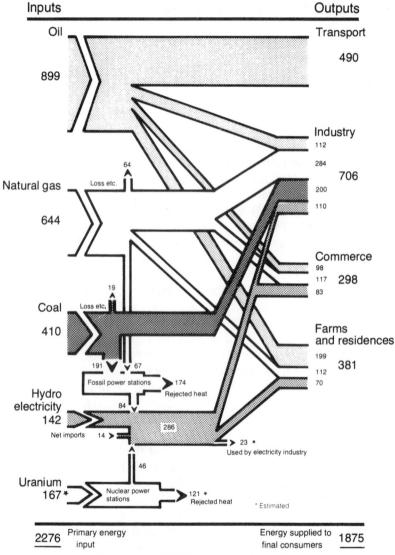

Units shown are 10^{12} British Thermal Units (multiply the
number by 293 to obtain the energies in millions of
kilowatt-hours).

TABLE 5
Ontario's primary energy consumption, 1978

	Consumption (natural units)	Consumption (10^{15} BTU)	Percentage of total consumption
Crude oil	2.2×10^8 barrels	1.3	48
Natural gas	670×10^9 cubic feet	0.67	26
Coal	16×10^6 tonnes	0.41	17
Hydroelectricity	38×10^9 kwh	0.13	5
Nuclear electricity	29×10^9 kwh	0.10	4

Source: adapted from Ontario Ministry of Energy, *Ontario Energy Review*, June 1979. Primary
electricity (hydraulic and nuclear) has been valued at 3412 BTU/kwh.

TABLE 6
Percentage of secondary* energy demand, Ontario, by fuel, 1965–2000

	1965	1975	1985 Current trends[†]	1985 Low energy use[‡]	2000 Current trends	2000 Low energy use
Gas	17	30	30	32	30	31
Oil	51	45	41	38	39	35
Coal	21	10	12	12	13	14
Electric			*			
Nuclear[§]	–	2	6	7	9	8
Hydro and thermal	11	13	11	11	9	11
Total Electric	11	15	17	18	18	19

* The energy consumed by refineries, pipelines, and electrical generating stations and that used
for non-energy purposes such as plastics, lubricating oil, and nylons are excluded from
secondary energy consumption.
† Current trends include a modest conservation programme.
‡ Low energy use includes more vigorous conservation reduction to approximately 2 per cent
per annum growth in demand.
§ Approximate percentage nuclear shares have been calculated by reducing all non-thermal
electric from 10,000 BTU/kWh to 3,412 BTU/kWh and recalculating the nuclear percentage.
Source: Ontario Ministry of Energy: *Ontario Energy Review* (Toronto, 1979), p. 42.

energy demand. In particular, nuclear energy comprises only about 4 per
cent of the total energy consumed in the province, not even a majority share
within electricity production. Any energy policy for Ontario must recognize
these hard facts and respond appropriately with specific policies to address

the present overwhelming dependence upon imported fossil fuels in a situation of ever increasing energy demands.

ENERGY POLICY

In 1973, a Ministry of Energy was created in Ontario. It was not, however, until April 1977 that the Ministry published a formal policy document, *Ontario's Energy Future* which addressed general energy policy matters.[3] Subsequently in 1979, a further policy document was issued, entitled *Energy Security for the Eighties: A Policy for Ontario*. Prior to these documents an understanding of Ontario's energy policies had to be pieced together from reading a wide range of ministerial speeches, government documents, and the observation of behaviour.

In the decade up to 1973, demand for energy in Ontario had grown by more than 5 per cent per year. As recently as the mid-1970s, figures produced by the Ministry of Energy, while not official projections, still indicated that on a 'business-as-usual' basis demand was expected to continue growing at about the same rate. By 1977, the rising price of energy began to affect demand and an economic slowdown began to be felt; the Ministry therefore identified a growth rate of about 3.5 per cent as being more realistic. Then in 1979, with Ontario's actual energy demand growing only by 2.5 per cent per year, policy targets were set to reduce the annual growth rate for all forms of energy to no more than 2 per cent and to achieve zero per capita growth in the demand for petroleum by 1985, without sacrificing economic growth.[4] How over-all growth rates of 5 or even 3.5 per cent were to be maintained was never made clear. However, the June 1979 *Ontario Energy Review* sets out details on current thinking about the nature and relative mix of future energy supplies (Table 6). A number of points emerge from this document:

1. On a 'business-as-usual' basis it is clear that Ontario, along with the rest of Canada, is likely to be essentially powered by fossil fuels well into the 1990s. Fossil fuels will still dominate energy supply in 1985 and the long lead times required to bring about major shifts in the pattern of energy use will ensure a similar energy supply pattern for the rest of the century. Only immediate, large-scale efforts made to alter the current trends could begin to change this pattern of development.

2. Electricity will still play a relatively minor role, measured in terms of equivalent BTU contribution, in 1985. But as with the federal policy, by far the largest projected energy growth sector is that of nuclear-generated electricity.

3. Even if the growth rate were reduced to about 2 per cent per annum, a reduction consistent with the objectives of Ontario's current energy conser-

FIGURE 9
Demand scenarios, Canada and Ontario

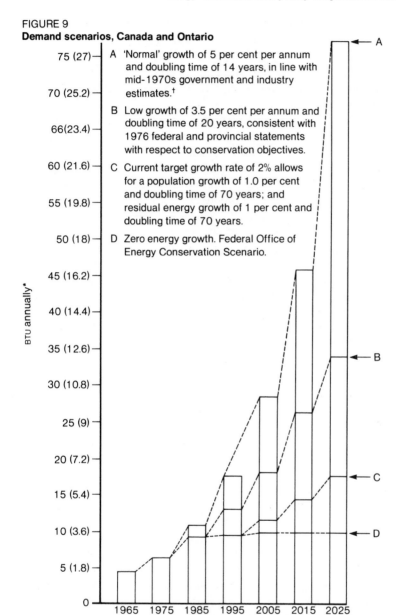

A 'Normal' growth of 5 per cent per annum and doubling time of 14 years, in line with mid-1970s government and industry estimates.†

B Low growth of 3.5 per cent per annum and doubling time of 20 years, consistent with 1976 federal and provincial statements with respect to conservation objectives.

C Current target growth rate of 2% allows for a population growth of 1.0 per cent and doubling time of 70 years; and residual energy growth of 1 per cent and doubling time of 70 years.

D Zero energy growth. Federal Office of Energy Conservation Scenario.

* Primary energy in 10^{15} BTU equivalent. Numbers in parentheses give approximate Ontario demand, taking Canadian demand as 3.3 times Ontario demand.

† Less than 1960s growth of 5.6 per cent per annum when 2025 demand would be 100×10^{15} BTU equivalent.

vation objectives, in 1995 the demand for energy would have increased by about 50 per cent over 1975 demands (Figure 9).

If we consider these three points together in the context of the national dilemma over expected oil shortfalls and an uncertain long-term natural gas future, we are surely forced to conclude that the energy future of Ontario seems precarious.

Oil is currently the most important form of energy consumed in Ontario. Most of this province's oil flows in pipelines from western Canada, principally Alberta and Saskatchewan, with a small amount of refined petroleum products entering the east of the province from Montreal refineries supplied by OPEC petroleum. In 1977 a pipeline was constructed across the province from Sarnia to Montreal in order to supply domestic oil to the market lying east of the Ottawa valley, thereby reducing the vulnerability of the eastern portion of Canada to interruption in its supply of OPEC petroleum. Unfortunately, the amount of oil available from the conventional western Canadian petroleum fields is now declining so rapidly that by the mid 1980s it is possible that there will not be adequate amounts of western petroleum to fill the pipeline. Fortunately the pipeline has been designed to pump oil in either direction and at that time it will be reversed to pump OPEC oil, if it is available, westward from Montreal into the heartland of Ontario. This particular scenario is even more likely in view of the announcement by the oil companies involved that it may well be as late as 1995 before a pipeline can be constructed to deliver any oil discovered in the Arctic.

Would Alberta tar sands oil be likely to alleviate this situation? A variety of constraints severely restrict the rate at which tar sands plants can be built and hence the likely flow of oil from them, and government reports conclude that in the foreseeable future the tar sands are likely to play at best a modest role in Canada's energy future and a relatively minor role in Ontario's energy future.[6] For example, in 1975 the Ontario government invested $100 million in the Syncrude tar sands plant of northern Alberta through the provincial Ontario Energy Corporation (OEC). This investment purchased a 5 per cent interest in the synthetic crude oil operation, or the equivalent of about 6,300 barrels a day. But the 1979 daily consumption of oil in Ontario was roughly one hundred times this figure, and in any case the OEC subsequently sold these shares. The increasing gap between projected growth in the demand for oil and ability to meet this demand from indigenous Canadian oil reserves will mean that sometime in the 1980s Ontario will be exposed inevitably to the uncertainties of the international oil market.[7]

Natural gas is evidently in adequate short-term supply. However, federal projections do not suggest any long-term optimism for natural gas supplies.

Growing western Canadian demand for gas and rising exports to the United States will eat into supplies available to Ontario. Moreover, increased eastern domestic consumption would require pipeline extension to the Maritimes, a costly investment not likely to be made in the face of an uncertain future supply. Finally, while there are many applications in all energy sectors where natural gas might replace oil, it would take perhaps ten to twenty years to bring about the changeover. Not only would this transition deplete natural gas reserves even more rapidly, but also the increased demand during the transition would occur at about the time when supplies of natural gas might well be running short.[8] High Arctic gas might have been thought to help in such a situation, and in 1975 the provincial cabinet approved in principle participation of up to $10 million in the Polar Gas project through the OEC. This project aims to tap natural gas located in the islands of Canada's far north and to connect it by a new Arctic pipeline with a terminus in Ontario. But that pipeline has yet to be approved let alone constructed.[9] Thus though Ontario's presently projected natural gas future is brighter than its oil future, it is still extremely problematic whether the natural gas share of the energy market can be substantially expanded and, if it is, whether it can be sustained for any substantial period of time (e.g., beyond 2000).

Coal imported from western Canada where very large reserves are known to exist could, in principle, play a more important role in Ontario's energy future in the industrial sector as well as in the generation of electricity. However, caution must be exercised about an expansion in the use of coal for several reasons. A main reason is that Ontario has essentially completed a thirty-year phasing out of coal-burning technology in the industrial sector. Reversing this trend, although possible, would require considerable private investment, take from ten to twenty years, and require the construction of an extensive and expensive coal delivery system from western Canada. Second, the coal-containing structures in western Canada are frequently complicated, and mining procedures must still be designed to permit the recovery of the largest possible proportion of the coal in such deposits, while meeting stringent safety and health standards. Third, coal mining is not a preferred occupation, particularly underground, and the lack of manpower is a constraint on production. Fourth, capital costs for mines are high and escalating. Lead times for the purchase of equipment and for bringing new properties into production are very long. Fifth, some of the producing provinces are not interested in the rapid development of coal properties for environmental reasons, and because it does not serve their immediate provincial objectives. And finally, burning coal in quantity involves serious air pollution problems for the consuming province, and possibly global climatic problems as well, unless expensive cleaning technology is added. Technology for the direct conversion of coal to gas or liquid fuels is under development, but at the present time these processes too are very expensive.[10] Despite the drawbacks, the resources are large and, in contrast to oil and natural gas, shortages are not an immediate problem. Therefore, coal may well prove a critical, if limited, bridge from our current dependence upon oil and natural gas to some other form of energy (see Part III).

Hydro-electric power provides a considerable energy resource in Ontario. In 1976 about 40 per cent of all electrical energy produced in the province was generated at hydro-electric installations, and as recently as 1960 nearly all electricity was hydraulic in origin. However, Ontario has harnessed most of the large, readily available, hydro-electric sites. About one-half of the untapped hydro power in the province is located in the far northwestern region on the Albany River and adjacent systems. The other half is in the form of individual small sites scattered throughout the province. Ontario Hydro's current expansion plans do not include any further hydraulic generating stations, though this exclusion is now under review. In fact, during the last twenty years a large number of small installations have been shut down in a steady movement towards large, centralized generating facilities.

A major public investment and energy growth area that is emphasized in *Ontario's Energy Future* and *Energy Security in the Eighties* is nuclear-generated electricity. In 1977 Ontario Hydro presented a Long-Range Forecast LRF48A (itself a scaled-down version of earlier expansion plans) which involved an enormous expansion in nuclear as well as coal-fired generating capacity. This expansion was necessary, in the opinion of Hydro's planners, to meet the projected demand, increasing by about 7 per cent per year, for electricity up to the year 2008. The 1977 total for all kinds of installed electrical capacity was about 22,000 megawatts; under the LRF48A plan, between 1977 and 2008 over 80,000 megawatts of additional nuclear capacity were to be installed (Figure 10) and a further 40,000 megawatts were planned from plants using fossil fuels.[11] The concept of an all-electric society was being seriously examined, with the bulk of the electricity to be generated in nuclear facilities.

These expansion plans have since been modified in a drastic way, in the light of recent projections of the demand for electricity which are substantially lower, being estimated in 1980 to increase by 3.4 per cent per year to the year 2000. However, the nature, costs, and implications of LRF48A are worth considering carefully as an example of what is involved in an attempt to move towards any major nuclear commitment. Even though Ontario Hydro's 1980 expansion plans call tentatively for only a further 3,400 megawatts of nuclear capacity (plus 600 megawatts of hydroelectric and 1000

FIGURE 10

Ontario East system capacity and peak demand scenarios
(Source: Adapted from Royal Commission on Electric Power Planning, Interim Report,
A Race against Time [Toronto, 1978], p. 32)

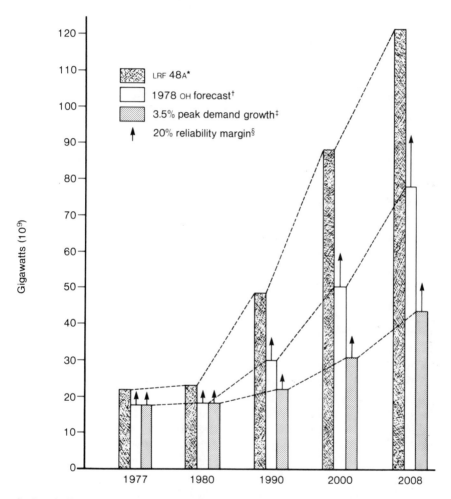

* Ontario Hydro, Long-Range Forecast 48A, 1975, for installed electrical capacity.
† Ontario Hydro, forecast of peak demand, 1978.
‡ Ontario Hydro, 1980 demand forecast is for growth at 3.4 per cent per year.
§ For purposes of illustrative matching of peak demand scenarios to required installed capacity
  scenarios we have added 20 per cent to peak demand figures to yield an installed capacity
  figure. Cf. 1977 figures.

megawatts of coal-fired capacity) by the year 2000 – these further 3,400 megawatts being in addition to the 8,600 megawatts expected from the nuclear facilities already under construction at Pickering, Bruce, and Darlington – it is important to recognize that, with the problems about the supply of fossil fuel, Ontario and federal energy policies both maintain that the only feasible large-scale replacement for fossil fuels in the foreseeable future in nuclear power. Thus there remains an important contradiction between government statements suggesting the need for a huge expansion of nuclear capacity to replace fossil fuels and the current slow growth (3.4 per cent) in the demand for electricity.

The percentage contribution which electricity will make by the year 2000 to total provincial energy requirements depends on two factors: the over-all rate of growth in the demand for all forms of energy and the rate of growth in electrical generating capacity. If we assume that the growth in total energy demand continues at the rate of 2 per cent per year until 2000 and that, contrary to current statements on energy planning, an expansion in electrical capacity similar to that proposed in LRF48A were to begin in 1980, then, by the year 2000, electricity would contribute about 40 per cent of the provincial demand for primary energy.[12] By the year 2020, if these trends continued, we would have reached the all-electric society, with two-thirds of Ontario's energy supplied from nuclear stations and most of the remainder from coal-fired installations. On the other hand, should total energy demand grow at 2 per cent and generating capacity expand at 3.4 per cent per year, as Ontario Hydro currently predicts, then in 2000 electricity would meet about 20 per cent of provincial demand and three-fifths of that would be from nuclear power. The all-electric age would then arrive a hundred years later, in 2100.

Thus a policy dilemma exists. Current expansion plans for the provincial electrical system appear inadequate in terms of the need for rapid substitution of fossil fuels. To replace fossil fuels in the next thirty years with electricity, a generating expansion program of the scale of the LRF48A projections is required, along with a growth in total energy demand of only 2 per cent per year. Such a growth rate is consistent either with a continuing depressed economy, a situation hardly conducive to generating the large sums required to finance both LRF48A and the conversion from a largely fossil-fuel economy to an electrical one, or on the other hand with a growing economy which pursues a very vigorous conservation program.

The contradictions between provincial energy policy and Ontario Hydro planning emerge in a comparison of the former's *Energy Security for the Eighties*[13] and the latter's 1980 proposals for a generation expansion program.

The former assumes that the following electrical capacity will be built by 1995 as a specific target for indigenous resource development:
1 2,000 megawatts of additional hydraulic generation
2 1,000 megawatts of lignite-fuelled capacity at Onakawana
3 8,600 megawatts of nuclear capacity at Pickering B, Bruce B, and Darlington.
Neither of the first two of these is included in Hydro's 1980 plans for its expansion to 1995. They are mentioned in Hydro's uncommitted program for 1996–2000, but the hydraulic expansion is expected to add only 600 megawatts. Hydro proposed the addition of nuclear generating capacity for the period to 2000 of 3,400 megawatts more than did the Ministry.

Our conclusion, then, is that Hydro's current, largely nuclear expansion program, to meet a 3.4 per cent per year increase in the demand for electrical energy is, in itself, no solution to the growing provincial problems over the supply of fossil fuel energy. On the other hand, a 'technical fix' in the form of a LRF48A scale of expansion has a number of formidable problems associated with it. One of these is its capital cost. Indeed, the financial requirements of LRF48A were such as could jeopardize the province's AAA credit rating in the international capital market – a possibility so alarming to Ontario politicians that it led to the formation of the Royal Commission on Electric Power Planning and a Select Committee of the Legislature to investigate Hydro's expansion plans.

Ontario Hydro had estimated in October 1975 that its expansion plans for the next ten years would require \$19 billion (constant 1976 dollars, Table 7). When it announced plans, as part of LRF48A, to install a further 29,000 megawatts of nuclear capacity in 1990–2000, the cost of that expansion was estimated to be over \$26 billion (1976). There are no published figures for the 39,000 megawatt expansion provisionally planned for 2000–08. These costs compare with a total capital expenditure by Hydro of \$7.9 billion (1976) dollars in the ten years before 1976. Such large and rapid increases in capital costs as are entailed by such programs as LRF48A present severe financing difficulties.

The Ministry of Treasury, Economics, and Intergovernmental Affairs has expressed its concern that Ontario Hydro, in undertaking such a program, would be competing for increasingly scarce capital funds in the pursuit of its capital expansion programme: 'In 1975, Ontario crossed an important threshold as borrowing increased by 400% in response to rapid increases in both Ontario and Hydro expenditure programmes. For the first time, combined public borrowing exceeded the amount that bond markets were readily willing to absorb.'[14] The Ministry's projections showed that Hydro's LRF48A capital requirements far exceeded publicly available capital (Figures 11 and

TABLE 7
Capital expenditures for long-range forecast 48A*

	Millions of current dollars	Millions of 1976 dollars[†]
1977	1,559	1,425
1978	1,623	1,367
1979	1,992	1,539
1980	2,323	1,648
1981	2,775	1,781
1982	3,367	2,037
1983	3,694	2,080
1984	3,952	2,079
1985	4,229	2,090
1986	4,786	2,230
1987–91	36,033	13,900
1992–96	58,484	16,467

* For Ontario Hydro's expansion programme LRF 48A see Figure 10.
† Current dollars adjusted for inflation using Ontario Hydro, *Generating Construction Index for Nuclear Plants*, Economic Forecasting Series (October 1977).
Source: Ontario Hydro, Comptroller's Division, *Review of the System Expansion and Financial Plans* (April 1977).

12). It is clear that both the short- and medium-term financial situations look extremely bleak, as a substantial financing gap is indicated between 1975 and 1982 and beyond 1982 net borrowing requirements and total availability of funds diverge so greatly that the former has not even been drawn.

The Ministry further warned that 'capital market analysis is a most intractable subject, influenced by so many intangibles that it is impossible at this stage to forecast with either precision or assurance.'[15] The capital availability forecast would have to be a gross underestimate before it could be assumed that no real problem actually exists. The Ministry's remark should serve as a reminder that Figure 12 may overestimate capital availability. In the event that the loans required over the next two decades could not be obtained – and the projections by the Ministry and Ontario Hydro are hardly encouraging – Ontario Hydro would have been obliged to seek substantial rate increases to fill the gap between revenues and expenditures.[16]

Another aspect of capital availability that merits attention is the time period for which capital is loaned. Rational planning of facilities, which have

FIGURE 11

Borrowing requirements and capital availability
(Source: Adapted from the submission of the Ministry of Treasury, Economic and Inter-
governmental Affairs to the Royal Commission on Electric Power Planning,
May 1976)

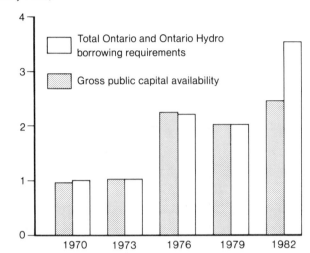

a long gestation period before they generate revenues, requires that most, if
not all, of the funds borrowed to finance the project should be long-term
loans at fixed interest rates. This allows more accurate cost calculations,
since the interest payments on the borrowed capital are known for the duration of the loan. However, Ontario Hydro reports an increasing proportion of
the money it has borrowed has been in short-term loans.[17] This trend shows
no sign of reversing.

Ontario Hydro was well aware of these concerns about capital availability.
In a review of its system expansion and financial plans, the Comptroller's
Division of Ontario Hydro projected a declining surplus of available funds
over borrowing requirements until 1980. Beyond 1980 a shortfall was projected that rose from a modest $37 million in 1981, exceeded $2 billion in
1990, and passed $3 billion in 1996. The Comptroller's Division summed up
the situation as follows: 'Perhaps the best interpretation of the result is that
the risk of a problem beginning in the early 1980's is relatively high if Hydro
proceeds according to current plans and policies.'[18]

In addition, more and more of Ontario Hydro's borrowing has been from
sources outside Canada.[19] The implications of this for Canada's balance of
payments have already been discussed. From the point of view of capital

FIGURE 12

Estimates of net public capital availability to Ontario and Ontario Hydro

(Source: Adapted from the submission of the Ministry of Treasury, Economic and Inter-
governmental Affairs to the Royal Commission on Electric Power Planning, May
1976)

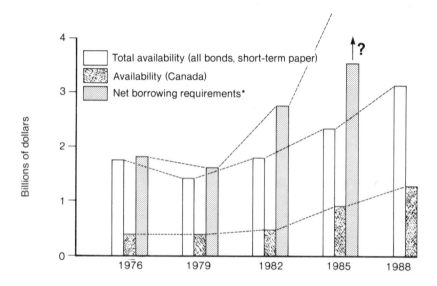

* In this case the discrepancy between net borrowing required and capital availability is so large
that the committee has evidently not attempted to extend the former projections beyond
1982.

availability, it means increasing reliance on the savings habits of foreign individuals and institutions. Governments in Canada have some control over
the level of savings within Canada for, within certain bounds, both fiscal and
monetary policies can be used to raise the level of savings. At the international level, however, the only way for Canada to attract increasing quantities of foreign funds would be to offer higher interest rates and to accept
shorter term loans. Both of these actions would not only increase the costs of
the capital projects so funded but, if taken to extremes, could also undermine Canada's international trading position. In summary any expansion of
the provincial electrical system on the scale of LRF48A would confront serious economic constraints and severely limit the province's ability to choose
and pursue other options.

BUSINESS AS USUAL

The continuing relevance of the Ministry of Energy's 1977 policy document, *Ontario's Energy Future*, was reaffirmed as 'the policy framework around which Ministry initiatives were developed' in the Annual Report of the Ministry for 1979.

The context for the discussion of energy policy developed in *Ontario's Energy Future* is the fossil fuel predicament in which Ontario finds itself, and the document makes it clear that the provincial commitment is to increasing use of fossil fuels: 'For some fifty years we will be primarily dependent for our energy supply upon coal, uranium, crude oil and natural gas from conventional and frontier sources, supplemented with synthetic oil and gas from the heavy oils and oil sands of western Canada and crude oil and possibly natural gas purchased on the world market.'[20] This plan for the future is completed with the addition of the commitment over the next seventy-five years to increasingly large amounts of nuclear-generated electricity. The commitment extends beyond the existing CANDU fission technology to 'advanced fuel cycle technology' such as plutonium and thorium exploitation that employ as yet unproven, indeed undeveloped, nuclear technology (see Chapter 5). Consistently, emphasis is placed on the expansion of the share of electrical energy, moving in the direction of the all-electrical society.[21]

The overall orientation of this policy document is to expand energy supply rather than to manage energy demand. Apart from the odd remark about a conserver society and a short section on energy conservation, the entire discussion centres on increasing conventional energy supplies. Indeed, the supply orientation of the document is strengthened with the remark: 'It is a fact that our food systems, health systems and lifestyle have been constructed upon the assumption of continuing and adequate supplies of energy. It follows that in the absence of adequate energy supplies our life support systems will become inoperative.'[22] This statement is open to a wide variety of interpretations: it can be read as a truism or as clearly false, since we could certainly reduce present energy consumption without ceasing to exist. What is disturbing is the attempt to back up a conventional supply orientation with scare tactics while avoiding a serious discussion either of limiting demand or of changing to renewable energy technologies.

In fact the economic emphasis in the document clearly falls on the further support of existing technologies and hence on existing financial interests in the energy supply industries. We are continually reminded of the economic consequences of not investing sufficiently in energy supply, in contexts where fossil and nuclear energy are taken for granted. At the same time the public management of energy demand is largely rejected and renewable energy technologies, such as solar and biomass technology, are either completely ignored or written off as irrelevant in a fifty-year time horizon; indeed, the document lists hydraulic and uranium derived energy as the province's only indigenous resources![23]

Ontario's most recent statement about energy policy, *Energy Security in the Eighties*, when read in conjunction with the *Ontario Energy Review* of June 1979, does introduce stronger themes of conservation and provincial self-reliance: it calls for a zero growth in provincial demand for oil by 1985, and an over-all growth rate of 2 per cent on average in total energy consumption by 1995, and at least 5 per cent of provincial energy needs in that year to be met from non-hydraulic, renewable, indigenous energy sources. The thrust of the Ministry's energy conservation policies is to encourage voluntary conservation by the citizens of Ontario: pursuing their enlightened self-interest, they will respond to the programs initiated, co-ordinated, and funded by the Ministry. Ontario, however, remains opposed to using the world price of oil as a benchmark for pricing Canadian oil products which would enhance the incentive for people to reduce their demand for oil; and at the same time no detailed program has yet appeared on how the target of 5 per cent renewable-resource energy supplies by 1995 might be reached.

Unfortunately, even a 2 per cent per year growth rate in Ontario's total demand for energy leads, by the year 2000, to a demand for energy that is about 50 per cent above what it is now. Indigenous resources can be seen as meeting less than 20 per cent of secondary demand in 2000 (when hydro and nuclear production are valued at 3412 BTU per kilowatt) and, moreover, the bulk of this new indigenous supply will be met by additions to the nuclear generating capacity. Indeed, *Energy Security in the Eighties* states: 'Ontario reaffirms its commitment to the safe and careful use of nuclear power to ensure a secure supply of electrical energy'; and further: 'Increasing Ontario's capacity for electrical self-sufficiency also means greater reliance on electricity as this is the energy form most easily produced from Ontario's own energy sources: conventional, renewable and recoverable.'[24]

Thus our conclusions about Ontario's current energy policy are essentially similar to those we came to about federal policy:
- There will be a growing dependence upon fossil fuels at a time of decreasing fossil fuel availability.
- There will be a rapidly growing dependence upon another non-renewable energy resource, uranium.
- The growth of highly centralized, capital-intensive technology in the hands of those with an existing economic interest in energy exploitation will be reinforced.

- Energy policy will continue to be focused on increasing the supply of energy, largely through investment in nuclear energy, even though the returns on the investment constitute, at best, only a partial response to the province's energy dilemma.
- There will be a relatively small commitment to addressing the demand side of the energy equation through an effective energy conservation programme.
- The social implications of energy policy will remain relatively neglected. We anticipate reinforcement of the trend towards a uniform industrialized lifestyle with decreased citizen responsibility, initiative, and participation. In short, the province's policy is one of energy insecurity.

4
Nuclear energy policy in Ontario

The technology for the controlled use of atomic or nuclear energy, first developed during the Second World War for the production of weapons and still used for that purpose, is playing an increasingly important role in contemporary energy policy. Atomic Energy of Canada Limited (AECL) has been the main nuclear research institution in this country, and Ontario Hydro has been the main institution in the country to apply nuclear technology to the generation of electricity. AECL provides a large part of the research and development involved in Hydro's nuclear development, which, in 1979, meant that 5,200 megawatts of nuclear capacity were in place in Ontario and another 8,500 megawatts capacity were under construction.

We have emphasized that, in the short to medium term (i.e., to about the end of the century), because of Ontario's large and growing dependency on fossil fuels, nuclear energy cannot be expected to play any but a modest role – about 10 per cent – in providing energy for the provincial economy. But in a longer term (i.e., to around 2025), its role could be quite different. Provided there is sufficient restraint on energy use and a commitment to expanding electrical output on the LRF48A level, these forty or fifty years could give the time, if not the money, required to allow the transition to an all-electric society. This transition, which would usher in profound changes in society, motivates the thinking and planning of many people.

All of us have a moderately good idea of what a 'business-as-usual' policy would be like for fossil fuels, but it is much harder to imagine and appreciate what might be involved in a nuclear future. Despite the major slow-down in Ontario's demand for energy as reflected in Hydro's planning, both the federal and provincial governments continue to support nuclear energy as the only viable replacement for fossil fuels, and nuclear investment is likely to dominate Ontario's capital borrowing for the next thirty years. However, many controversies still surround nuclear energy and many misconceptions still prevail.

THE COMMITMENT TO NUCLEAR ENERGY

There is no doubt that a major commitment to nuclear energy exists in many government and private institutions throughout the country. The massive nuclear programme that is included in Ontario Hydro's plans cannot be seriously launched without substantial government and industrial backing. The programme has been proposed despite growing domestic and international controversy over the economic, medical, and political consequences of nuclear power. Furthermore, a variety of recent official and semi-official statements have indicated that in the long run this province, and Canada, are moving towards an all-electrical society and that electricity would account for perhaps 90 per cent of our energy requirements by the year 2050. At the present time, no non-nuclear source of energy is known that would be capable of generating the huge amounts of electricity required for such an all-electric society. A commitment to an all-electric society at this time is, therefore, a commitment to nuclear power.

The full extent of the provincial commitment to nuclear energy only clearly emerged, however, in the 1977 Ministry of Energy document *Ontario's Energy Future*. As that document makes clear, the Ministry of Energy contemplates a long-term commitment to nuclear energy, a commitment not simply to the existing and relatively straightforward CANDU fission processes, but to yet-to-be-developed 'advanced fuel cycle' technology. The message is clear; a greatly intensified nuclear programme for the province is envisaged that will involve not only many more nuclear reactors but also new and much more complicated nuclear technologies.[1]

Such a commitment is massive and necessarily long-term, as some considerations emphasize. For instance, nuclear reactors have an expected lifetime of around thirty years. According to Ontario Hydro's LRF48A expansion programme (see Figure 10), new nuclear plants were going to be installed right up to the year 2008. Thus, even if nuclear expansion were to be halted instantly at 2008, the nuclear system would continue to operate until roughly 2040 or about 60 years from today. In the 2000–08 period, however, the nuclear system was projected to expand by an unprecedented amount. By then it would hardly be economically or socially feasible suddenly to cease commissioning new plants, leaving thousands of skilled workers without jobs and moving against the vested nuclear interests of what will certainly be some of the largest, technically most expert institutions in the province.[2] Moreover, advanced fuel cycle technology cannot be ready for commercial application much before 2000. Its advent would then coincide with mounting fissile wastes made economically very attractive by their further exploitation through such technology, probably with declining uranium reserves, and certainly with rising uranium prices, failing fossil fuels supplies, and rising energy demand. In this situation, transition to advanced fuel cycle technology would be well nigh impossible to resist, at least if supply-economic considerations dominate. But even to be economically attractive this technology would have to be employed for at least the thirty-year lifetime of one reactor. To exploit fully the large additional quantities of nuclear energy available, its use would need to continue well beyond that time.

Any reasonable investigation results, therefore, in the conclusion that, once made, the commitment will carry into the long-term future and on a large and unprecedented scale. This larger commitment might have been surmised from the sheer economic costs of the presently planned commitment to existing CANDU technology as well as the theoretical opportunities for much greater exploitation of the uranium resources that advanced fuel cycle technology promises. But it is not a commitment the nature and implications of which have heretofore been publicly discussed with the people of Ontario.

It is clear furthermore that strong support exists at the federal level for nuclear energy. It is impossible to ignore the substantial support AECL has enjoyed during the past thirty years, the existence of federal policies and subsidies for encouraging the construction of nuclear plants, and the arguments put forward in the federal energy strategy and LEAP documents.[3] A 1977 federal study on disposal of radioactive wastes took an equally positive view of this, as yet unsolved, problem.[4] Overall, federal policy seems optimistic about the resolution of the environmental and safety concerns often expressed concerning nuclear energy. At a public meeting in 1978, G. MacNabb of Energy, Mines and Resources, strongly defended the expansion of the nuclear system, arguing for it on grounds of reliability and cheapness and praising Ontario Hydro for being willing to 'take a gamble' with Atomic Energy of Canada Limited.[5]

Finally, a large industry has grown up in Canada around the nuclear power system. This industry has organized itself into a number of trade organizations – the Canadian Nuclear Association and the Association of Electrical and Electronics Manufacturers – that have been active in their support of nuclear power.

THE NUCLEAR CONTROVERSY

Although there is clearly considerable commitment to the development of an extensive nuclear power system in Canada, an important controversy has developed over both the feasibility and the desirability of undertaking such a programme. For example, there is increasing public concern about the safety of nuclear plants, the security of potentially dangerous material, the problems associated with the reprocessing or disposal of radioactive wastes, and the social and the ethical implications of a commitment to this particular form of energy. In the United States this controversy has led to what has been termed a 'virtual moratorium' on the construction of additional nuclear power plants. A number of bodies in Australia, the United States, Sweden, and Britain have been set up to explore the issue, and in 1975 the government of Ontario appointed the Royal Commission on Electric Power Planning to inquire into the long-range planning of the province's electrical power system.[6] Headed by Dr Arthur Porter, this commission identified the nuclear issue as one of the key questions that must be examined in its overall investigation. The Commission's interim report on this issue arrived at conclusions largely in agreement with our own.[7]

While grave concern over the safety, health, and environmental consequences and risks of pursuing nuclear energy is justifiable, it must be stressed that to see these as the only public concerns is to seriously misrepresent the issue. Just such a distortion has been the unhappy consequence of the course of much of the current debate. In addition, these concerns are well documented elsewhere.[8] To correct the present imbalance in the debate, our discussion focuses on the effectiveness, and on the economic and social consequences, of pursuing the present policy on nuclear energy.

Nuclear energy: a sufficient technology?
The view widely held both in government and private circles is that nuclear energy will play a major, if not totally sufficient, role in meeting growing energy demands as fossil fuels run short. However, a very rapid and extremely costly expansion of the nuclear generating capacity would be required to move from our present position as a society powered by fossil fuels to one powered by nuclear-generated electricity.[9] The growth rate involved would be far beyond the current projected expansion of our electrical system. Moreover, to convert the existing capital equipment in the industrial, commercial, and residential sectors to electricity from their present dependence upon oil and natural gas would be costly and time-consuming and create special pro-

blems in the transportation sector. Similarly, great strains would be placed on the capital, skilled manpower, and other resources necessary to expand the nuclear-electric system. While these constraints could no doubt be overcome with sufficient time and the application of sufficient resources, the question remains whether such a large commitment to an inflexible technology really constitutes the best use of society's resources.

But if nuclear power is not the universal panacea for energy problems as it is often portrayed, why does it play such a central role in the minds of decision-makers? There appear to be a number of reasons. First, fully aware of the risks of depending upon imported energy resources for Ontario's economic well-being, it is not surprising that the government is committed to the development of secure, indigenous uranium energy resources. Second, nuclear power has had the image of being the energy form of the future, a panacea for our energy woes by virtue of its being 'clean, cheap and inexhaustible.' Although recent experience has shown that none of these adjectives is fully accurate, continued statements to this effect by government and industry reinforce this image, thereby leading to continued support for the technology. Third, the nature of nuclear technology is directly amenable to centralized control and represents a straightforward extension of existing technical, planning, and control procedures.

These latter two factors are key elements of the 'technical fix' approach to public policy that is so widespread today. Within this mental framework, serious consideration is not often given to the somewhat different technical and economic structures of decentralized, renewable energy technologies or of a major governmental involvement in demand management. This is the way we understand stated premises of public policy such as 'apart from hydraulic sources for electrical generation, uranium is the only indigenous primary energy source found in significant quantities in Ontario' and 'electricity is, consequently, the only energy form relative to which the Government of Ontario can design policies with a minimum requirement for considering the decision-making power of other governments.'[10]

Moreover, centralized control and increasing economic importance provide incentive for private capital and companies to enter the market. In the early 1970s, when electricity demand was still growing at least 3 per cent annually faster than gross national product, it would take only twenty years to double electricity's share of the economy, and gain control over a fundamental economic factor at the same time. In this connection Simon Rippon, editor of *Nuclear Engineering International*, remarked: 'The big industrial concerns have not entered the business for quick profits ... For the most part

TABLE 8
Uranium commitments and recoverable resources, Canada and Ontario
(in thousands of tonnes)

	Canada		Ontario	
A. URANIUM COMMITMENTS				
Projected nuclear capacity by 2000 (megawatts)	82,000	60,000	47,000	31,000
Cumulative consumption to 2000	95.0	80.0	70	56
30-year consumption to 2030	345.0	254.0	200	131
Existing export commitments	73.4	73.4	62	62
Total	513.4	407.4	332	249
B. URANIUM RESOURCES RECOVERABLE*				
Measured	83		49	
Indicated	99		76	
Inferred	307		246	
Total	489		371	
Adjusted reserve†	377		282	

* At up to $156 per kilogram.
† The sum of tonnages in the measured, indicated, and inferred categories weighted with factors of 1.0, 0.8, and 0.7 respectively to reflect reliability.
Source: Royal Commission on Electric Power Planning, Interim Report, *A Race Against Time* (Toronto 1978), p. 142.

the position of industry is that ... for the wellbeing of their company they must establish a foothold in this sector of the business.'[11]

Another facet of the 'nuclear energy myth' is that the technology is often touted as being mature, available, and ready for deployment. In fact, however, existing technology is relatively inefficient in its use of uranium and uranium supplies are not assured (Table 8).[12] To avoid another energy crisis, this time in terms of uranium supplies, new nuclear technology must be developed immediately for deployment in the 1990s. Research and development for the new 'advanced fuel cycle technology' involving the use of plutonium and thorium is projected to cost $2 billion and to require twenty years of effort. As well, the technology to store and dispose of wastes, or to dismantle outdated reactors, has not been developed. Clearly, from a technical point of view, a mature nuclear technology is by no means in hand.

A further factor in the promotion of nuclear energy is the very heavy institutional commitment that Ontario Hydro, the nuclear industry, and the scientific establishment now have in its development. This commitment is reinforced by the strong personal commitments often found among the personnel of such institutions and in government. A very large staff at Ontario Hydro is dependent for its future existence upon an expanding nuclear energy programme. Moreover, it is the nature of future nuclear technology that it is really only controllable by the same technological elite who are supported by its expansion. In addition, considerable government prestige is already invested in the nuclear power programme. A major change in policy emphasis away from nuclear power to other forms of energy might well be construed as an admission of error and be a cause of political or electoral recriminations.

Much of the case for nuclear energy rests on the claim that it is a clean, cheap, and inexhaustible energy source. This discussion has cast doubt on these claims, suggesting that they be critically examined. This we now do.

A clean energy form?

Is nuclear power a 'clean non-polluting' form of energy? It is in comparison with fossil-fueled power stations and the air pollution problems associated with their production of vast amounts of sulphur dioxide, particulates, nitric oxide, carbon dioxide, and in some instances radioactive materials in the fly ash. In a nuclear power plant, where the source of energy lies not in chemical combustion but in splitting the nuclei of unstable uranium atoms, no air pollution of a traditional chemical sort results. However, the nuclear fission process produces radioactive pollution. Normally, the vast majority of the radioactive byproducts of the nuclear fission process are contained within the power plant. But even during normal operating conditions small amounts of nuclear wastes are released under controlled conditions in the form of gases to the environment or in very dilute form into the cooling water required for the operation of the plant.[13]

Further, at some time, the nuclear fuel required to operate the power plant must be removed from the nuclear reactor and replaced with fresh fuel. The spent nuclear fuel contains very large amounts of highly dangerous radioactive materials that must be separated from the biosphere for long periods – in some instances in excess of 100,000 years. To prevent them from causing harm, the management of these nuclear wastes presents one of the major challenges to the acceptability of nuclear power for society.[14]

Significantly, many countries that have a substantial investment in nuclear generating plants have had a long history of 'irregular occurrences.' These range from the unintended opening of a radioactive liquid waste pipe and pulling the wrong fuse, thereby shutting down a reactor and damaging the turbine (both by Ontario Hydro) to a major control room fire, and two 'critically out-of-control' reactor cores (both in the United States).[15] As a result, in many countries generating stations run well below projected capacity. Canada has been comparatively fortunate to date, though the track record of different plants has varied. However, Canada's experience of nuclear power currently extends to a relatively small number of commercial plants in their early years of operation.

We find it predictable, but alarming, that investigations of major failures consistently reveal such thoroughly human causes as ignorance, carelessness, rivalry, and greed, while the defenders of nuclear safety consistently quote calculations based on non-human engineering failures. It is impressive to be told, on the basis of an extensive scientific study, that the chances of a major nuclear accident are less than 1 in ten million.[16] But the actual accidents occurring have by and large not been caused by the kinds of engineering failures to which such reports have been almost wholly restricted. In the 1975 accident at Brown's Ferry, the simple misuse of a candle by an ill-trained worker eventually knocked out all seven of the official safety systems. A more serious outcome was only prevented by the fortuitous linkage of another, different part of the plant's water supply to the reactor core. A chief safety systems designer for the plant evidently resigned in dismay. In the April 1979 accident at Three Mile Island, a major explosion threatened, radioactive gas was emitted to the surroundings, a partial evacuation was ordered, and a major evacuation of over a million was seriously considered. It seems that the incident arose because in a particular unforeseen circumstance there was only one safety backup system and it was erroneously turned off manually by an ill-trained employee. Moreover, the engineering calculations themselves in the safety studies have been subject to fundamental criticism, even excluding the human elements.[17] The conclusion can only be that the public should treat such promises of safety extremely cautiously at the present time.

If we make the dubious assumption that the risk of nuclear accident is small, a further important consideration is whether people really want to accept even a small risk of a catastrophic accident by developing this technology if there is a viable alternative. Consider this parallel: would you purchase a vacuum cleaner if there were a very small, but finite, risk of it exploding and taking the neighbourhood with it, when there were alternatives? And there are alternatives to nuclear technology (see Part III).

The uncertain costs of nuclear energy

Now let us consider briefly the 'cheap and inexhaustible' part of the claim for the superiority of nuclear energy. With respect to inexhaustibility for CANDU, the issue may be quickly settled. Table 8, taken from the Interim Report on Nuclear Power in Ontario of the Royal Commission on Electric Power Planning, clearly indicates that contemplated nuclear developments already tax the limits of Canadian uranium reserves without additional export commitments. And it is not obvious that pressure to export uranium can be resisted, either politically or economically. For these reasons nuclear power advocates look to the development of advanced fuel cycle technology in the longer term. But this transition represents a very substantial technological transformation, and it may not be an ultimate panacea. For example, plutonium recycling may provide a 30 per cent decrease in the rate of resource consumption, but the net extension of uranium supplies will be shorter than this if the system continues to grow. While transition to the thorium-burning cycle could substantially increase the effective lifetime of supplies, and perhaps by enough to warrant the label inexhaustible, we are here entering the realm of large uncertainties where wholly undeveloped but complex and exacting technologies are assumed. Instead of further speculation, we shall consider economic costs.

The economic arguments are quite simple and hinge on the comparatively low operating costs of large nuclear generating plants. However, cost-advantage claims are subject to many more uncertainties than are usually recognized and are sufficiently large to throw into doubt the long-term economic competitiveness of nuclear-generated electricity. If these doubts can be substantiated, much more serious attention should be paid to non-nuclear, especially renewable, energy technologies as economically reasonable energy policy responses to current problems.

Remember that post-2000 nuclear expansion is assumed to rest increasingly upon advanced fuel cycle technology because of decreasing uranium fuel supplies and that this technology has yet to be developed and tested. It is, therefore, entirely unclear at the present time what the costs of such plants might be. In particular, given the substantial increase in complexity involved in the transition to this new technology, it is not at all clear that these plants will produce electricity at a price comparable to the claims now made for existing CANDU plants.[18]

Furthermore, the figures for existing CANDU-type plants are subject to substantial uncertainty (Table 9). Although certainly desirable, detailed cost comparisons between CANDU and many other energy technologies are not possible. There are as yet too few data available on alternative energy technologies to be able to make these comparisons. There are, however, data

TABLE 9
Capital costs of future generation capacity

	Capacity (megawatts)	Project cost, Ontario Hydro ($ million)		Percentage change
		1974	1975	
Pickering 'B'	4 × 516	1,239	2,019	+63
Bruce 'B'	4 × 769	1,881	2,870	+53
Darlington	4 × 850	2,074	3,461	+62
Thunder Bay	300	170	261	+54
West System	600	512	684	+34

Source: Ontario Hydro, Submission to the Ontario Energy Board, *Bulk Power Rates for 1976* (May 1975), Vol. II, Table 3, p. 8.

available on conventional fossil-fueled electrical generating plants and Ontario Hydro has used them as the basis of a cost comparison with nuclear-generated electricity. This comparison can be used to demonstrate the major uncertainties relating to the costs of nuclear energy. The arguments here are strong enough to open the prospect of similar arguments being made for comparisons with alternative energy forms.

According to data published by Ontario Hydro the non-fuel costs of a 750 megawatt CANDU plant are more than twice those of a coal-fired plant of equal size.[19] Most of these are capital costs. When fuel costs are included, coal-fired and CANDU plants are equally costly when the average annual load factor is 40 per cent. At a 60 per cent load factor, the CANDU plant is 20 per cent cheaper than the coal plant; at an 80 per cent average annual load factor CANDU is 31 per cent cheaper. For smaller generating plants, the cost advantage of CANDU is somewhat reduced and for larger plants it is somewhat increased, according to Ontario Hydro's calculations.

These cost comparisons are, of course, only estimates. A decade or more from now, the capital and operating and maintenance costs of these two types of generating plants could be significantly different from the estimates on which today's expenditure programmes are being based. Indeed, the range of uncertainty is large enough that coal-fired plants might turn out to be as cheap as or cheaper than nuclear plants, just as now appears to be the case in the United States.[20]

For example, the nuclear/coal cost advantage would be reversed, at an 80 per cent load factor, if the non-fuel costs of nuclear plants with a 750 megawatt unit size were 100 per cent higher than was estimated by Ontario Hydro.

At a load factor of 60 per cent, the non-fuel costs need only be underestimated by 50 per cent for coal plants to be cheaper than nuclear ones. Errors of from 50 to 100 per cent in estimates of capital costs are quite possible. Ontario Hydro's submission to the Ontario Energy Board reported increases in project costs of this magnitude over one year. It is indeed conceivable, therefore, that non-fuel costs of coal and nuclear generating stations could be underestimated by 100 per cent or more over the approximately twelve to thirteen years that Ontario Hydro allows to bring a generating station into service.

The promise of larger cost savings from the construction of much larger units than the 850 megawatt units now entering service must be regarded with scepticism; CANDU units of this size have scarcely been designed, let alone built. Moreover, it is arguable that the repeated construction of identical units may offer a better means of securing production economies than unduly rapid attempts to achieve what may turn out to be elusive economies of scale. Ontario Hydro had envisioned 1,200 and 2,000 megawatt stations in its LRF48A forecast.

A further indication of the uncertainty surrounding the estimates of capital cost for electrical energy supply arises out of a comparison between the capital costs for Ontario Hydro's expansion programme cited above and those made by the federal government. All the data are brought together in Table 10, which reveals that the federal government projected capital expenditures on electrical energy as a percentage of gross national product to be 33 per cent higher than the equivalent estimates made by the Ontario authorities. Without more detailed information about the underlying assumptions that led to these estimates, it is not possible to be completely certain that the two sets of estimates are inconsistent. Nevertheless, there are several reasons for believing that these estimates are irreconcilable, and that it is the provincial estimates that are too low rather than the federal estimates that are too high.

For example, we can dispense with the possibility that the different percentages reflect different projected rates of economic growth rather than different projected costs of electricity generation. The federal projection of growth in GNP during the 1975–90 period exceeds the Ontario projection of growth in gross provincial product (GPP) for the same period. Other things being equal, this would tend to produce lower capital costs for electrical generation as a percentage of GNP than similar costs for Ontario as a percentage of GPP. In fact, as the table shows, the reverse situation is projected.

Second, as already stated, the federal estimates reported here are the lower of two sets of projections. The alternative 'low-price scenario' in which energy prices are assumed to rise less rapidly than in the high-price scenario

TABLE 10

Comparison of investments in electrical energy supply, Canada and Ontario, 1976–90

	1976–80	1981–85	1986–90	Total
Capital requirements for electric power ($ billions)				
Canada (EMR)*	21.7	32.5	37.0	91.2
Ontario (Hydro)†	6.8	9.1	12.0	27.9
GNP (EMR)*	912.1	1,116.2	1,438.5	3,512.6
GPP (Treasury)‡	373.6	471.5	582.3	1,427.4
Electrical energy investment as percentage of				
GNP (row 1/row 3)	2.38	2.91	2.57	2.60
GPP (row 2/row 4)	1.82	1.93	2.06	1.95

* Federal estimates from Table 2.

† Estimate by Ontario Hydro, Table 7. 1976 dollars in Table 7 were changed to 1975 dollars using the 11 per cent increase in construction costs reported in Ontario Hydro's *Economic Forecasting Series*, 1976. A capital expenditure estimate for 1976 was obtained from Exhibit B92, submitted to the Select Committee on Ontario Hydro by Ontario Hydro.

‡ 1975 gross provincial product (GPP) obtained from Statistics Canada, User Advisory Services, May 1976. Projected growth rates from Ministry of Treasury, Economics and Inter-Governmental Affairs, Submission to the Royal Commission on Electric Power Planning, May 1976, p. 20.

These figures were drawn up before the reduction in forecasts of electrical demand.

involves some 43 per cent higher capital requirements for electric power between 1976 and 1990 than the figures in the table. In other words, the federal figures being compared with the estimates for Ontario are presented in the federal study as the lower of two reasonably possible futures for electrical energy in Canada.

Third, it is well known that Ontario leads the provinces in nuclear power stations. Therefore, the long-range plan for Ontario in which nuclear capacity was to be added at twice the rate of fossil-fuel capacity represents a more nuclear-intensive programme than the overall Canadian programme. Bearing in mind that nuclear power facilities are much more capital intensive than fossil-fuel facilities of equal capacity, it is to be expected that Ontario's generation expansion programme would appear more, rather than less, expensive as a proportion of GPP when compared with the federal programme as a proportion of GNP. Once again, this is reason to suspect

Ontario's estimates of being too low rather than the federal estimates of being too large.

Finally, it must be recognized that in recent years estimates of capital costs for large projects of all kinds have typically proven to be much too low. Only some of these errors are attributable to inflation rates in excess of those anticipated. In many cases, real construction costs, especially of large-scale projects involving new technology, have proved far more costly than were expected. Possibly these underestimates are made by bureaucrats who are likely to approve projects that appear relatively inexpensive and then find it impossible to cancel them once commitments have been made. Certainly some types of skilled labour and raw materials have presented bottlenecks to capital investment on a broad scale and the costs of these inputs have risen disproportionately, thus raising the costs of the projects. If this trend of underestimates continues into the future, then both the provincial and federal projections of capital costs for electricity generation may prove too low.[21]

Fuelling costs must also be considered. It is commonly thought that the fuelling costs of nuclear plants are such a small proportion of the total costs that the relative economic attractiveness of nuclear power is virtually unaffected by the costs of nuclear fuel. An examination of Ontario Hydro's own data disproves this belief. Nuclear fuel costs would have to be only three times greater than Ontario Hydro is anticipating for 750 megawatt nuclear plants, operating at an 80 per cent load factor, to lose their advantage over coal-fired plants. At a 60 per cent load factor, the nuclear fuel costs need only double beyond those currently projected for the cost difference between nuclear plants and coal-fired plants to be eliminated.

The possibility that Ontario Hydro is underestimating the long-term fuelling costs of its nuclear plants is just as strong as the possibility of errors in estimating the capital costs. Ontario Hydro's forecasts for world prices of nuclear fuel (Table 11) reflect the view that there will be tight world uranium supplies for the next few years and that they will then ease considerably and permanently until at least the year 2015. This view is not shared by the majority of experts who appeared before the Select Committee on Hydro Affairs Investigating Ontario Hydro's Uranium Contracts with Dennison Mines and Preston Mines.[22] The Ontario Ministry of Energy's own consultant on uranium availability and prices, David S. Robertson, forecasts a significant scarcity of uranium in the 1990s with the price of uranium oxide reaching $100 per pound in current dollars because 'in the Western world at large, it remains difficult to imagine uranium find rates and building of production facilities such as to meet presently perceived demand even

TABLE 11
Change from previous year in
world price of nuclear fuel

	Percentage change
1975	25.0
1976	30.0
1977	35.0
1978	25.0
1979	20.0
1980	10.0
1981	8.0
1982	7.5
1983	7.0
1984	6.5
1985–2015	6.3

Source: Submission to the Select Committee of
the Legislative Assembly Inquiring into
Hydro's Bulk Power Rates for 1976.

with the significant reductions in demand of the last few years and the significant expansion of ore reserves created by price movement.'[23]

Because of this sort of forecast, most commentators favoured Ontario Hydro's 1978 uranium contracts with Dennison Mines and Preston Mines, since they provide some protection against rising world prices. The main features of these contracts are that together they apply for forty years, during which time 200 million pounds of uranium will be sold to Ontario Hydro. In any year the price will be a percentage (50 per cent for Dennison Mines and 33⅓ per cent for Preston Mines) of the difference between the prevailing world price and a base price consisting of production costs, depreciation, mining taxes, and $5 per pound (updated) 'fair' profit. These contracts will provide sufficient uranium for Ontario Hydro's existing and committed plants throughout their expected lives, which means, according to current plans, for all plants up to and including the Darlington generating station. Further contracts will have to be signed to ensure adequate fuel supplies for all the plants that succeed Darlington. From the testimony provided to the Select Committee by experts, it is doubtful that new uranium discoveries of sufficient magnitude, plus adequately expanded production and processing facilities, will in fact allow Ontario Hydro to fuel all of its expanded nuclear capacity from indigenous Ontario reserves should it attempt expansion on the scale of LRF48A. Such circumstances will lend great pressure to the moves already afoot in some circles to begin reprocessing spent fuel and to introduce the thorium fuel cycle.

In the light of these considerations, which suggest considerable long-term increases in uranium prices, it is significant that Ontario Hydro's forecast of uranium prices is higher than that used in its comparison between nuclear and coal-fired generating stations: '[The validity of the cost estimates] depends on Ontario Hydro getting its uranium at prices that are shielded from the full effect of spiralling prices. These are caused by the severe imbalance between supply and demand for uranium expected to develop within the next few years and last beyond the protection of our current contracts which cover most of our requirements through 1979.'[24]

While Ontario Hydro has obtained some protection for its existing and committed plants, the degree of protection is unlikely to reach the 40 to 50 per cent assumed by Ontario Hydro in its coal/nuclear cost comparison.[25] There is at present no indication of the extent, if any, to which Ontario Hydro will be able to protect itself from rising world prices for uranium in order to fuel the plants proposed for the 1990s and after.

Another important factor to be considered in connection with the future price of uranium is the formation of a uranium 'cartel.' Although there has been considerable dispute as to whether it was indeed a cartel or merely an 'informal marketing arrangement,' it is generally believed that the price of uranium surpassed 'anything that the marketing arrangements had contemplated.'[26] What is significant here is that not only have some of the principal suppliers of uranium already come together to co-ordinate their marketing efforts, but also that the price of uranium has increased *ten* times as a result of 'natural market forces' and that the marketing arrangements played no part in this. It is immaterial whether or not the producers concerned are at present actively working together. The point is that they have already demonstrated a willingness to do so. We are reminded that OPEC was formed in 1960 but had little impact on world oil prices until 1973. On the basis of present information, the geographical distribution of economically exploitable deposits of world uranium is no less concentrated than that of oil.[27] Thus it would be unreasonable to deny the possibility that an active uranium cartel will be formed in the future.

In order to gain a further indication of this possibility, it is necessary to consider the capacity of buyers and sellers of uranium to respond to increasing prices. On the supply side, if moderate price increases could be expected to lead to large increments in reserves owing to the extra profits to be gained from exploration and development, then this would act as a brake on the escalation of uranium prices. Though this sort of situation might prevail in

Australia, where recent large discoveries have been made, it likely will not occur either in Canada or Europe. It is impossible to say at this stage how these complicated international circumstances will be reflected in prices. On the demand side, one thing that distinguishes nuclear power stations from fossil-fuel stations is that it is far too expensive to build fuel options into nuclear plants to enable them to run alternatively on fossil fuels. While fossil-fuel plants can be built to run on a variety of fossil fuels at little extra cost, once a nuclear plant is built, it must run on nuclear fuel for the thirty-odd years of its operation. Nuclear plants without price protection will have to pay the asking price for uranium or shut down.

Moreover, indigenous supplies of raw material do not necessarily offer real protection against high world prices to domestic buyers. This is obvious from experience with the price of Alberta oil, which has risen steadily towards the world price. One of the reasons for this is that a policy designed to keep down the domestic price of an internationally traded commodity tends to undermine domestic investment incentives – the case argued by oil companies operating in Canada. Short of nationalization or extremely costly investment subsidies, both policies that may be politically unacceptable, there is little governments can do to prevent uranium mining capital and expertise from looking elsewhere in response to a government-imposed differential between domestic and world prices of uranium.[28]

Thus the individual uncertainties in both capital and fuelling costs for CANDU-type nuclear power stations are each individually sufficiently large to undermine the claimed nuclear/coal cost advantage. When taken in combination, these individual uncertainties could each be considerably smaller and yet the total cost advantage of nuclear over coal would still be undermined.

This discussion does not lead us to conclude that coal-fired plants will prove to be as economical as nuclear plants. There might be an unforeseen increase in the price of coal, especially if Ontario Hydro switches from U.S. to western coal, and capital expenditures for coal-fired plants could rise if more stringent standards for sulphur oxide discharges were imposed. But there are also countervailing considerations. There is a real possibility that the costs of disposing of radioactive wastes and of decommissioning nuclear plants could be much higher than anticipated, especially since at present there is no entirely satisfactory way of performing these tasks, regardless of cost. Coal-fired electric generation may thus turn out to be less expensive.

That this can be suggested seriously serves to illustrate the major uncertainties about the economic costs of obtaining electricity from nuclear energy. It is these uncertainties that concern us, especially since nuclear power receives much of its support from those who believe it to be a proven technology capable of providing energy, if not at prices too cheap to meter (a 1950s claim), at least at prices substantially cheaper than the alternatives (a 1970s claim). As we have shown, the nuclear programme in Ontario was based on stations of increasing size that at some stage would probably require reprocessed fuel and/or the use of thorium – both unproven technologies. In addition to these technological uncertainties there are those surrounding the capital costs of nuclear plants, the availability of capital, and the price of uranium.

THE CHOICE OF ENERGY POLICY

The Ontario Ministry of Energy places high priority on security of energy supply, by which is understood a combination of four factors: freedom from susceptibility to external manipulation; economic viability, especially with respect to the international market place; future flexibility; and supply adequate to meet demand.

Only indigenous energy resources can be free from external influence. In Ontario only hydraulic and nuclear-generated electricity are thought to be both indigenous and available, though the former is considered essentially fully exploited. However, the remaining indigenous resource, nuclear power, has been held to be economically competitive with its rivals and hence to be economically viable. Moreover, it has been held to be an essentially unlimited resource and hence to be capable of meeting future energy demands.

We have argued that the case for the economic viability of nuclear power is a great deal less certain than has been supposed. Ironically, one of the uncertainties associated with nuclear power is precisely its sensitivity to internationally driven uranium prices. Thus it is not entirely free from external international pressures. Nuclear power may also turn out not to be a wholly indigenous energy source. One reason for passing to advanced fuel cycle technology is to avoid exhausting Ontario's indigenous uranium reserves; but then it is far from clear that this will not cause a substantial increase in our international vulnerability in the highly sophisticated technological fields involved. Moreover, despite appearances, even with the large expenditure contemplated and the optimistic development programme faultlessly realized, only roughly 10 per cent of expected secondary energy demand will be met in 2000 by nuclear power. (See Table 6).

Finally, the magnitude of the investment required, together with the inherent nature of the technology, leads us to believe that nuclear energy lacks future flexibility. Commitments to highly centralized technologies are hard to reverse since an entire spatial pattern of energy use grows around them. Strong economic interests are then vested in its perpetuation. The sheer magnitude of the resource investments required to expand the nuclear system is very likely to prevent any real possibility of development of alternative energy technologies. This fear is one strongly shared by Ontario's Ministry of Energy, yet the present emphasis on nuclear expansion is exactly the commitment most likely to foreclose on future flexibility of choice.

Thus it is far from clear that nuclear-generated electricity completely satisfies any of the features of a desirable energy source. Other energy sources – which the Ministry of Energy has chosen to discount – are the indigenous energy resources found in solar and biomass energy in particular and to a lesser extent in wind energy and geothermal energy accessible via heat pumps. These energy sources are also indefinitely renewable and are unowned or largely publicly owned and hence not subject to the same potential threat of international price manipulation as is uranium. Technologies to exploit these resources, unlike nuclear technology, are 'small scale' and largely controllable within the provincial borders. Because there are multiple diverse technologies available or capable of development, there is a high degree of future flexibility of choice inherent in this type of energy exploitation. A strong case can therefore be made that these technologies satisfy three of the four factors constituting a desirable energy source.

The question of the economic viability of renewable energy technologies is a large one, depending not just on their cost but upon the larger structure of the energy economy of a society and its 'lifestyle.' At the moment, only preliminary and often incomplete information is available. Nonetheless, the studies drawn upon in Part III lead to the conclusion that such alternative energy policies are likely to be as economically viable as the present nuclear policy in the medium term and considerably more viable in the long term. The reasons for believing this lie in a relatively simple technology that may be improved relatively easily and is based on a truly renewable resource. There are, as well, still other reasons for caution in regard to an intensive nuclear energy policy.

Present policy calls for a massive nuclear bureaucracy. Knowing what we do about bureaucracies, for how long can we reasonably expect this bureaucracy to operate without serious error? What has happened when errors have been made? The inevitable tendency is for the nuclear bureaucracy and the government supporting it to cover up 'irregular occurrences' and to use inaccessible technical arguments to justify policy decisions.

We take as example a small, local incident: the unusually high radon concentrations in the air around Port Hope unquestionably involved increased cancer risks – albeit, not large increases.[29] Yet in the Port Hope situation, government experts evidently did not discuss the matter in straightforward terms. Instead they seemed to know of only two relevant figures: the measured radon activity in the air expressed in units meaningful only to the initiated; and the legal exposure limit for uranium miners. They apparently failed to specify that these limits were recommended for occupational exposure and not for children. Nor was it clear from the official announcements that exposures smaller than the recommended limit still carry a not inconsiderable risk. No doubt officials did not want to cause a public panic, so the public were kept more or less in ignorance 'in their own interest.' In this reasonable manner people are denied participation and their social responsibility is eroded.[30]

What chiefly concerns us here are the political implications of what happened. The critical point about nuclear power is that its potential for centralizing bureaucratic control is truly awesome in comparison with that of many other energy sources. First, because it is a complex and risky technology it requires strict, centralized management. In an all-nuclear future not one person in ten thousand will have any knowledge of, or participation in, the energy system that will shape every aspect of his or her life. Second, the technology poses international and domestic military risks. A commitment to nuclear energy will mean both acceptance of increasing pressure for domestic nuclear growth and involvement in the international spread of nuclear technology and fissile material.[31] Canada would make a worthier international contribution – and perhaps as much or more money – by developing low-cost, flexible renewable energy technologies for export. Nuclear fuels and nuclear plants are also potential targets for sabotage – which has not been included in calculations of nuclear accident risks, another reason for public caution. Nuclear facilities may eventually have to be policed by a special force, which would require special powers of surveillance and emergency control and hence further erode the institutional bases for freedom in our society.[32] In the case of nuclear power especially, the potential for restricting public responsibility and freedom is awesome. Such potential is a part of the social consequences of energy policy, a dimension that is of fundamental importance and our focus in Chapter 5.

CONCLUSION

The nuclear programme is monopolizing a large share of Ontario's capital borrowing capacity; it is also absorbing a large fraction of Ontario's skilled manpower that might otherwise develop and apply its abilities to other energy technologies. Even so, nuclear electricity is at best only a partial response to the energy dilemma facing Ontario, except in the very long run. And associated with a large-scale commitment to nuclear power are a number of pressing environmental and social issues that have not yet been resolved to the satisfaction of many diverse groups within society. There is a serious danger that the very large amounts of resources required to pursue the nuclear option may not be forthcoming. Ontario would then be left with all of the negative social consequences *and* a half-developed programme that would leave it in an even worse energy predicament than it presently faces.

The question therefore arises whether or not there are other policy initiatives which might not address Ontario's energy dilemma more rapidly and at equal or less cost, and with fewer difficult and unresolved environmental and social problems. We do not wish to be seen only as critics of nuclear technology. Many of the considerations that have been raised, and that will be raised in Chapter 5, remain true of an all-electric, coal-powered society. Such a society would still be characterized by centralized bureaucratic control, by massive capital investments for the profit of a few, and probably by high environmental risks. Our arguments are for movement in another institutional and technological direction. In the meantime, we must deal with the present commitment to nuclear energy.

We do not argue for the abandonment of the existing nuclear system; only for not adding to it. Indeed, the electrical system that will be in place when existing commitments are completed will form an important part of our conventional 'bridge' into our non-conventional energy future. Nor do we argue for the abandonment of research on nuclear power generation. Further research may lead to an uncontemplated technological breakthrough, and we must pursue research on safe radioactive waste disposal and 'decommissioning' of nuclear generating stations and uranium mines, because we already have operating nuclear facilities. In any case having technical access to nuclear power, if it is ever needed, is a part of maintaining maximum future flexibility of choice. What we do argue for is a much more vigorous pursuit of alternative energy technologies and the structural adjustments that must accompany them. Correspondingly, there must be a reduction in the nuclear research budget and a corresponding increase in renewable energy research budgets.[33]

5
Energy policy as social policy

The preceding chapters emphasized that energy policy is deeply connected to social policies, to the style and quality of life we enjoy and hence to our culture, and to the ways in which we think of ourselves as a people. Energy policy is, therefore, a fundamental social policy. Most of us, however, are still not accustomed to considering the social implications of energy policy in a systematic way. As long as sufficient energy is available to satisfy demand and provided that energy can be generated in a way that does not impose undue risks or hardship, we can presumably turn our attention to other matters. To be sure, it is widely recognized now that energy policy has wider social implications, relating to, for example, a nation's economic position or the health effects of energy production. Such implications obviously have to be taken into account when energy policy is designed, and we expect government to take appropriate regulatory action where necessary. But that is generally the extent of our thinking.

Energy, in its various physical forms, is usually conceived of as merely one more commodity whose production is regulated by market forces and fairly rigid needs. Moreover, it is regarded as a commodity that is so essential in our society that any harmful side-effects in its production are a small price we all should be willing to pay. Today this attitude is becoming increasingly difficult to maintain. As the scale of our energy projects increases and as we are forced to utilize increasingly sophisticated means of satisfying the demand for energy, the social effects of our energy supply systems become ever more noticeable.

In addition, it is also becoming increasingly clear that the traditional representation of our energy situation is quite misleading. Discussions on energy technology usually imply that, subject to modest conservation targets, energy demands are given and reflect a lifestyle that policy-makers regard as unalterable and/or sacrosanct; that today there exist only a small number of technically and economically feasible ways of generating energy, and these are given objectively, free of value judgments, and that we are already making full use of these technologies. But in response it has been repeatedly argued that the relation between energy use and lifestyle is complicated, and improved lifestyles can be achieved with lower energy demands. Further there are, in fact, alternative sources of energy that are now untapped, not because of any deep technical or economic problems, but because they do not fit our present preconceptions of the ideal energy source.

If such alternatives exist, we are obviously not committed by economic or technological imperatives to continue to rely solely on our present energy systems. Economics and technology do, of course, impose constraints on how we as a society can regulate our energy supply, but they can never determine what we ought to do within these constraints. It is not as if technologies are somehow beyond our influence. The technological attractiveness of a scheme does not necessarily imply its overall superiority; there may be other, equally feasible technologies that are preferable on other than technical grounds. Nor in the past have major technological obstacles prevented the successful development of a host of technologies – take, for example, the twenty-five year nurturing of nuclear technology. The prospect of future financial gain is usually sufficient to ensure the rapid solution of a large number of, though not necessarily all, technological problems.

Present energy technology research and development is typically directed towards applications that fit in with the existing industrial and institutional structure of our society. But the institutional structure is not absolute, just as the present state of our technological capabilities is not a fact of nature but the result, very largely, of past technological policies and investment decisions. Similarly, present technology research and development policies will greatly influence the technological possibilities open to the next generation. Thus it would seem wise not to bias such policies too much towards existing institutional interests, unless we consciously and publicly approve of them. Otherwise, we may find that we have unnecessarily constrained our future technological, and social, options.

One of the central points we want to emphasize is that the objective constraints on energy policy-making in fact leave us a considerable amount of freedom, at least at the present time. How exactly we shall satisfy our energy needs, within the constraints set by economics and technology, depends ultimately on the kind of society we want. Therefore, such social considerations are not a minor, subsidiary theme, to be dealt with when all 'important' economic and technical decisions have already been made; rather, they are the very quintessence of energy policy-making.

We have identified at least some of the main social factors related to energy production and supply and divided them into three categories: those having to do with energy availability, with energy production, and with energy planning.

SOCIAL POLICY AND ENERGY AVAILABILITY

The quantity of energy available, its quality – for example, cleanliness, ease of transportation, and thermodynamic effectiveness – and its price have all had their impact on our society, and it is probable that they will continue to. For example, the availability of cheap liquid fossil fuels has encouraged the rapid spread of private transportation, which in turn has had a tremendous influence on our land use patterns and made possible large-scale urbanization and industrialized agriculture. Cheap transportation has enhanced the tendency towards centralized commodity production and increased inter-regional dependencies.[1]

Cheap and abundantly available fuels strongly influence industrial processes, and recycling of energy 'wastes' becomes uneconomical and consequently pollution loads increase. The pulp and paper industry, which is notorious for its organic pollutants and electricity consumption, could be very much less polluting and virtually energy self-sufficient if it used its own wastes to generate electricity – a transition now slowly taking place in Ontario. But because energy is amply available and cheap our industries tend to be exceedingly wasteful. Similarly, the availability of cheap energy has led to the construction of buildings that are extremely energy inefficient, often using three times the space heating and cooling energy of their well-designed counterparts. Office buildings are not infrequently heated on one side and cooled on the other, and temperature levels in buildings are often unnecessarily hot in winter and cold in summer. Low-cost fuel has also led to the use of motor vehicles with low thermodynamic efficiency and to the favouring of energy-intensive transportation modes such as motor vehicles and planes over lower energy use alternatives such as trains, boats and bicycles.

A characteristic of our energy economy as a whole is its tendency towards spatial centralization. Energy fuels are most often not consumed at the site of initial acquisition, but are typically transported to a small number of refineries or electric power plants where they are 'upgraded' to equivalent high-temperature energy and subsequently redistributed. Correspondingly, industries have tended to concentrate their activity into increasingly large, centralized industrial forms. The spatial centralization of energy-producing industries (oil and gas companies, Ontario Hydro's electrical system) is partly attributed to the spatially concentrated, high-grade energy forms we have been exploiting; partly to the economies of scale achievable with large, centralized technologies; and partly to the concentration of ownership in the energy supply sector. The tendency to centralize among energy-consuming industries arises from a combination of cheap, plentiful, and already central-ized energy and the concentration of ownership. Especially for the almost universal regressive rate structures, particularly for electricity and natural gas, where the cost of energy to the buyer decreases with increasing con-sumption, it has in the past been profitable to substitute energy and capital (in the form of machines) for labour. Under these conditions, it has also usually proved more profitable than not to employ both larger and less energy efficient processes. Some of the most important social dimensions of energy policy arise from the consequences of these centralization tendencies.

First, the substitution of energy and capital for labour in industry may create unemployment. According to Hannon, this process, exacerbated by population growth, has been creating a labour surplus of around 7 per cent per year over the past twenty years.[2] The North American economies have hitherto been able to expand demand fast enough to absorb much of this surplus but this is no longer the case, especially as energy is becoming increasingly expensive and more is being imported. Even the huge govern-ment deficits characteristic of recent years now seem inadequate to take up the slack in the private sector. Moreover, large-scale, centralized energy technologies themselves have proved highly capital intensive and are becom-ing increasingly so. Not only do the vast sums invested generate relatively little direct employment, but they are diverted from other, more labour-intensive uses where they would generate higher employment. Further-more, the kinds of jobs directly generated by nuclear technology tend to be concentrated in the heavy construction and professional sectors and at a few plant locations. By contrast, the more decentralized solar technology would generate jobs in the small engineering and trades sectors and tend to be more dispersed throughout the community.[3] Both the level of employment and economic and spatial employment are important social concerns.

Second, the centralization of the energy economy has been a major factor in the regional concentration of economic activity and hence in the develop-ment of national and provincial disparities. Nationally, for example, Alberta has been treated as a resource hinterland by Ontario and Quebec just like any 'underdeveloped country'. At present it is exploited for food and fossil fuels.

But now the economic situation is reversing itself and more aggressive steps are being taken to attempt the development of an indigenous manufacturing sector in Alberta to support employment after fossil fuel supplies decline. Provincially, similar regional asymmetries have developed within Ontario between the Windsor-Toronto-Ottawa strip and the vast northern portion comprising 80 per cent or more of the province. These regional economic and social disparities have become a major economic concern to the provincial government.[4] Such disparities may be seen even within large cities, where cheap energy has made extensive urban transportation possible and with that the modern residential sprawl combined with urban centralization of work. But these spatial asymmetries raise some of the most important social issues of our time: equality of opportunity, environmental quality, the future of culture hitherto supported by city cores, transportation accident rates, and so on. These spatial patterns would be significantly modified if, as expected, energy prices continue to increase. But they would eventually be radically changed if the energy economy were decentralized through large-scale shifts to decentralized, renewable energy sources.

Third, the centralization of the energy supply system has itself been an important factor in encouraging a proliferating demand for energy. It is characteristic of centralized systems that they can provide freedom of choice of development to a region essentially only by oversupplying it with energy at any given moment. This has in fact been Ontario Hydro's policy. By contrast, a decentralized system can be adapted to local development decisions, largely at the time they are actually made. Correlatively, it is attractive to a centralized system to have energy-intensive heavy industries that consume large quantities. The fluctuating demands of small users can then be handled as additions to a heavy basic demand. But this has led historically to large consumers with lower prices – that is, a regressive electrical rate structure – with the result that high energy use is encouraged. With rapidly mounting costs for new generation capacity, small consumers in effect subsidize the large. Centralization also causes thermodynamic inefficiency to increase faster than the supply of energy itself. This is, in part, because energy is being brought from increasingly distant sources and in part because losses from the redistribution system increase roughly as the square of the distance covered. By 1985, for example, it is expected that Ontario Hydro will need as much as 10 per cent of its primary energy demand for its own operation and this figure does not include the internal energy demands of its municipal utilities and similar drains that belong in calculations of efficiency. In a more decentralized system, where energy supply can be chosen to meet local requirements, the importance of these three factors can be substantially reduced and, therefore, so also can the pressure towards increasing energy consumption.

Furthermore, a centralized system is subject to risk of large-scale costly failure. For example, on 2 April 1978 a fuse costing a few cents was wrongly removed, shutting down of one of the 750 megawatt reactors at Pickering, Ontario; the resulting damage to the steam generation turbine totalled $3.5 million. But for the fortuitous presence of a spare turbine at the site, the plant would have been out of operation for a long period, substantially reducing the Pickering performance rating as well. Irregular occurrences of such large scale would be impossible in a thoroughly decentralized energy system.

Another consequence of the centralization of the energy economy is the neglect of renewable energy forms. Renewable forms of energy are in general only locally exploitable because transportation over longer distances presents difficulties. Such energy sources (sun, wind, hydro power, geothermal, and biomass conversion) are consequently neglected altogether or the energy is transformed into electrical energy that can be transported. The high cost of electricity grids and the need for their tight control does not, however, encourage the exploitation of small sources of renewable energy, unless the energy is to be used locally. Hence our present predominant reliance on non-renewable energy sources plus the few large hydro-electric sources available.

Finally, one of the major motivations for centralization of the energy sector is the desire to achieve tighter control. While this is lauded in the sphere of private enterprise, it is a central theme of western democracy to fear its consequences in government. Yet in the prevailing energy economy awesome decision-making power is concentrated in relatively few hands.[5] This is particularly true of the nuclear system, which is not only too complex to be easily understood, but also has built-in rationale for secrecy based on the safety and security implications of its technology. Since energy decisions are socially important, their further removal from public control constitutes a major concentration of political power and a major dimunition of public control.

The consequences of centralization were not forced upon us by the technological and economic exigencies of our energy supply system; rather, we have hitherto tacitly chosen to allow them to develop. Such policies are, however, open to re-evaluation precisely because the fundamental social consequences they engender are themselves subject to political choice.

The final aspect of energy availability that we want to consider in this connection is the future commitment that our present consumption patterns imply. Such commitments have, it seems, become increasingly significant with the rapid increases in the centralization and the scale of our energy system. Power stations, pipelines, large mines, and oil wells all require large investments that discourage the early abandonment of such projects, irrespective of rising economic, health, and environmental costs. Perhaps even more important, many changes in social organisation made possible by the wide availability of relatively cheap energy cannot be changed in a short time span. Land use patterns, the industrial base, and the design and construction of our buildings, to mention only the most obvious ones, are all likely to influence our energy requirements far into the future.

In particular, the future commitments nuclear power forces upon us should be taken very seriously. Nuclear technology takes a long time to develop and deploy to economic advantage. Nuclear wastes remain dangerous for a quarter of a million years. They have to be guarded or, if deposited in some geological formation, monitored continuously. The plutonium nuclear reactors produce has only two known uses: to power yet more reactors or to construct nuclear bombs. When a large fraction of our electricity is generated in nuclear plants, even a serious accident or a fundamental re-evaluation of the hazards of low-level radiation might well be discounted when either one ought, perhaps, to be sufficient cause to cancel the nuclear programme.

The conclusion drawn from all these considerations must surely be that we can ill afford energy policies that do not allow for future flexibility of choice. There are in fact two reasons why present energy policy decisions are crucial to future energy options. The first is the relative irreversibility of the decisions because of the sheer magnitude of the economic and social investment involved. The second is the importance of declining fossil supplies, which can be used to 'bridge' our economy into a different energy future. They can be used only once for this purpose and probably not at all if we do not act decisively.

Both of these factors were recognized by the Ontario Ministry of Energy's 1977 policy document.[6] Yet there is good reason to fear that the policies advocated in that document will indeed lock Ontario inflexibly into specific and costly fossil and particularly nuclear energy technologies far into the future. Moreover, in the post-2000 era the rapidly rising costs of the conventional fossil/nuclear programme and the existing economic and institutional investment the province will have in it will surely make serious pursuit of alternatives impossible. Thus if serious research and development devoted to alternative, non-nuclear renewable energy technology is postponed to that time, there is a real danger that it will become a self-fulfilling prophecy that there is only one feasible course of energy technology development. Yet decentralized, multiple source, non-nuclear renewable energy technologies offer prospects of a much more flexible and more secure energy system were they to be pursued now, while our fossil fuel bridge to the future still exists. Part III offers support for these claims and for the social importance of alternative technologies.

ENERGY PRODUCTION AND RISK

The social consequences of energy production are equally as important as spatial considerations. For example, although conditions have improved, coal miners are subject to lung disease. Earlier than normal retirement is often still necessary. Similar health hazards are found in fossil-fuel generating plants and refineries and all working in them may be subject to substances potentially harmful to the respiratory system. Improvements in working conditions have been fought for bitterly, and the struggle for a healthier working environment continues. Similarly, fossil-fuel generating plants have in the past spilled their waste products into the air of all those living near them and into the surrounding environment generally. The accumulated health and environmental effects of fossil-fuel energy technologies are important negative features of their use.

The biological effects of nuclear energy are perhaps the least clearly understood. The International Commission for Radiation Protection (ICRP) has, somewhat hesitantly, recommended radiation exposure limits for the public at large as well as for workers in the nuclear industry. Workers in power plants and uranium mines are legally allowed to receive doses many times larger than members of the general public. It is, however, not often appreciated that these limits represent a tradeoff between negative health effects, and the risk of cancer in particular, and expected social benefits of nuclear energy. There does not appear to be a threshold below which no deleterious effects will occur. Indeed, the ICRP was quite explicit when it recommended the present limits, stressing that the assessment was preliminary and scientifically premature, and that it involved 'a compromise between deleterious effects and social benefits.'[7] One might, therefore, expect that these 'allowable' limits would be treated with some respect, and that as few people as possible would be exposed to radiation doses as high as the legal maximum. This is decidedly not the case. The dismal story of the lack of safety precautions in the uranium mining industry is well known.[8]

Less well known are the hazardous working conditions in nuclear plants. Not infrequently, many plant employees have accumulated their maximum radiation dose and unexposed outside workers have to be hired. In the United States transient workers already comprise a significant fraction of the nuclear industry's labour force. These workers are virtually untrained, and they may receive minimal instructions about safety procedures and the risks they will run. If skilled workers are required the industry often faces considerable difficulties in finding enough unexposed workers. (In Canada, Atomic Energy of Canada Limited is known to have enlisted the help of large numbers of armed forces members to clean up reactor spills.) The nature of this situation would become particularly serious in a large and expanding nuclear program where within a few years the nuclear industry might 'run into serious road-blocks due to a lack of available maintenance personnel.' Quality control becomes exceedingly difficult when it takes many men, often inexperienced and poorly instructed, to do one job.[9]

Apart from the increased chance of serious accidents as a result of this lack of labour quality control, the fact is that the nuclear industry is exposing a considerable number of people to worrisome levels of radiation. Ionizing radiation causes genetic defects that manifest themselves in the descendents of exposed individuals. The exact measure of the risk is hard to determine, however, because of the large numbers of animals required for experimentation. It is important to note that it is 'largely immaterial whether the defective genes are introduced into the gene pool by a few individuals who have received large doses of radiation, or by many individuals in whom smaller doses have produced correspondingly few mutations.'[10] The nuclear industry is therefore playing 'nuclear roulette.'[11] Again, it is significant that problems such as this are not usually considered when energy policies are designed.

There is, of course, also the risk of serious nuclear accidents,[12] and we remind the reader that all of these considerations apply as well in the case of military nuclear research and development, from which public scrutiny of safety levels is excluded.

Of course all energy systems have deleterious health and environmental consequences associated with them. This includes renewable energy technologies as well as those of coal and nuclear energy. But until recently no serious attempt was made to compare likely accident rates among energy technologies or likely environmental effects among them. The task is made the more difficult by the absence of any large-scale extended experience with renewable energy technologies.

Intuitively, one expects most accidents in the renewable technology field to be largely mechanical, small-scale, and non-lethal and environmental damage to be low. A recent study by Inhaber suggests that in fact renewable energy technologies might have higher accident rates associated with them than nuclear technology.[13] Inhaber's results have been severely criticized, and the data base is as yet of dubious reliability. We concede, however, that there will be accidents in the renewable technology fields, and that they may even be substantial in comparison with 'normal' nuclear accident rates – excluding major reactor accidents. The crucial issues will then be the *kinds* of biological damage inflicted, the spread of the damage across the population, and the ability of individuals to control their own well-being through safer work habits. Choosing the quality of risks incurred is an important ethical and social choice.

If the ready availability of energy and its production have widespread social consequences in the foregoing categories, it is also true that the energy system generally is one of the main pillars of the existing social order. Energy production and delivery provide many hundreds of thousands of jobs and determine the affluence level of entire provinces. They also have created some of the largest and most influential economic institutions in the nation. Thus support of a given energy system is support of an entire set of social arrangements and social power that extends from the energy institutions to all with whom they deal. Correspondingly, to seek to change the energy production system may be to seek important changes in the existing social structure.

ENERGY POLICY-MAKING AND PLANNING

Policy and planning have traditionally been dominated by technical and economic considerations. Consequently, these tasks have been the prerogative of a relatively small elite of technocrats, and important energy decisions have often been made without any public debate. Indeed, this undemocratic practice is often justified by appealing to the sophistication of modern energy technology and financial management. Whatever the sophistication of the technology – and exaggeration is easy – the implications of energy decisions are usually very easy to state. Yet more often than not experts do not translate their findings into generally understandable language.

This anti-democratic consequence is unhappily all too natural an outcome of the capital- and technology-intensive energy system we are developing and it holds in other, similarly structured institutions. Often it can develop so gently and with such evident good intention that its real social significance is easily overlooked. Consider the following statement by R. M. Dillon, deputy pro-secretary, Natural Resources Secretariate:

A prime function of the Commission [Royal Commission on Electric Power Planning] will be to establish what the need is for electricity in the Province and the role that nuclear energy should play. The need for electricity must be expressed not only in technical terms – percentage increases per year, historic demand trends, etc. – but must also be translated into social terms – the implications of deliberate limitations of supply for instance – so that people can visualize what it means to them. We are convinced that people will be willing to listen with an open mind to explanations of what the nuclear programme is all about provided they are convinced of the need.[14]

The quote reveals Mr. Dillon's assumptions that rising provincial energy consumption is inevitable and that nuclear power is either the only or the most desirable way of meeting the increased demand. The policy-making ideas implicit in this passage are that an objective need for electrical energy can be established, that establishing such needs is most appropriately the responsibility of (para-)government agencies, such as the royal commission or the load forecasting branch of Ontario Hydro, as in the past, and that the responsibility of government is then to explain the findings to the electorate and, if necessary, help to convince the electorate that the optimal decision has been made.

Enough has been said to indicate the depth of error involved in assuming an objective need for electricity. Provincial electrical demand reflects choices of technology, transportation, housing, packaging, and so on and reveals a broad set of sociopolitical and lifestyle commitments. Parts III and IV reinforce this conclusion. For just the reason that demand does reflect broad sociopolitical policy Mr Dillon's insistence that the issues be translated into socially relevant terms is praiseworthy but, as his remarks show, the process of disenfranchisement can be subtle and relentless under large-scale centralization.

This is no isolated example. The Ministry of Energy's policy document *Ontario's Energy Future* illustrates this approach to policy-making throughout. The stated purpose of the document is 'to advise the public of our energy future,' but the document presents only one energy 'future.' That future is presented not as a path to be chosen, but as a prediction of a foregone conclusion. For example, 'between now and the end of the century an increasing amount of the energy used in Ontario *will be* in the form of electricity. An increasing amount of the energy used in electricity *will* be produced through nuclear generation because there is no real option.'[15] Even the title of the document is in the singular. No substantial discussion of social issues is offered; the emphasis is on technical and economic considerations. Finally, energy futures based on non-nuclear, renewable technologies are not given a hearing, since 'it is the opinion of informed people in Canada and the United States that renewable energy sources will contribute little to satisfying our energy requirements through the end of this century.'[16] Instead, the reader is informed of the promise and attractiveness of advanced, centralized nuclear technologies without being told what is involved, on the grounds presumably that the details are incomprehensible and/or may be safely left to the experts involved.

With increasing scale and centralization, the trend in all our institutions, and energy institutions in particular, seems to be towards decreasing public control. Single projects are of such magnitude and so much is at stake financially that the temptation to underplay potential problems must be almost irresistible. In the particular case of nuclear energy, safety and sabotage considerations add to the impetus for strict control. The result might easily be the creation, 'in the citizens' own interests,' of a nuclear priesthood and the corresponding further denial of participation of the ordinary citizen. Fittingly, it is now the nuclear engineer who invites us to strike a 'Faustian' bargain, to choose between dependence in luxury and freedom in the cold.[17]

Any sophisticated technology that also fulfils a socially important role creates the danger of elitism. This does not by itself condemn the use of such a technology, it merely points to the necessity of keeping a watchful eye on those in control. For in their preoccupation with what appear to be the narrowly technical requirements, they may in fact be creating a socially undesirable situation. Control may be more difficult than it would appear, especially once the institution takes on the vast scale of modern energy institutions. Ontario Hydro, for example, is a crown corporation, responsible to government; yet one has to judge that one of the primary factors motivating the long parade of commissions, boards, and committees that have examined Ontario Hydro's activities over recent years is the difficulty government has in getting a solid grip on what goes on inside it and what the repercussions of its policies are likely to be. A former minister of energy, James Taylor, remarked on his resignation in January 1978 that Ontario Hydro 'is a cabinet Minister's nightmare and the government has little control over it.'

This dilemma is of course false, for the road towards the centralized, fossil/nuclear society is not the only one that will allow us to satisfy our energy requirements. But to seek instead to develop a more decentralized, smaller scale energy system through renewable energy technologies is to seek to rearrange substantially the existing energy-related economic/employment structure. More important, it is to seek to diffuse highly concentrated, social and economic power by offering local communities more autonomy and political control.

It would be ironical if, at a time when energy is starting to be seen as a fundamental social concern, energy policy-making became progressively removed from public control. At a time when our decisions in the energy field represent increasingly irrevocable commitments to a future social order, it is essential that energy policy-making be considered the responsibility of all citizens. This, however, clearly requires a substantial transformation of our energy institutions, for these we have inherited from an era in which energy was regarded as merely a cheap marketable commodity. Again it should be stressed that our present energy policies tend to ignore both the primacy of social considerations and the existence of fundamental conflicts of interest. And as long as social considerations are predicated upon technological and economic 'imperatives' and disputes are thought to arise merely from ignorance, we shall continue along our present energy path, stumbling from one crisis to the next.

PART III
AN ALTERNATIVE ENERGY POLICY

6

An alternative energy strategy

Though some modest changes have begun to appear, Canadian energy policy remains committed to conventional patterns of energy production and use. We believe that this energy strategy is by no means the only practicable one. A growing body of evidence argues the existence of a feasible and, we believe, a highly desirable class of alternative energy strategies open to Canada.

The business-as-usual energy policy currently offered us accepts as given that the future will bring large increases in the demand for energy. It places emphasis on energy supply measures, with continuing dependence on fossil fuels and a growing dependence on nuclear energy as the only feasible strategy open to our society. From these assumptions follows the necessity of developing our Arctic resources, the tar sands, and Arctic pipelines and the inevitable transition to a predominantly nuclear-electric society.

The objective of the next two chapters is to outline and discuss the feasibility of an alternative energy strategy that will result in a more socially satisfactory and secure energy future for our society than is likely to emerge from the energy policy currently being pursued by our policy-makers. The strategy is put forward as being appropriate for Canada as a whole and is illustrated, in some detail, by considering its importance for energy problems in Ontario. Our proposed plan is more satisfactory than current approaches because it specifically addresses as a whole the environmental, social, economic, and political concerns that comprise the energy dilemma. The strategy is more secure because energy requirements will largely be met by indigenous, renewable – hence inexhaustible – energy resources through the use of technology and expertise well within Canada's industrial capacity.

The major components of this alternative strategy are:
- a major commitment to increased efficiency in the use of energy, achieved through adopting a variety of conservation and innovative technological

measures, with the objective of reducing the growth in demand and in the long term reducing the total consumption of energy in Canada;
- the harnessing of renewable energy resources at a rate and on a scale sufficient to meet a significant proportion of energy requirements over the next twenty-five years, and the majority of these requirements after some fifty years;
- the implementation of 'intermediate-term' policies for remaining fossil-fuel reserves and existing nuclear generating capacity to facilitate the transition from predominantly fossil-fueled economy to one largely based upon renewable energy resources.[1]

Although much neglected until recently, there is now hardly room to doubt the enormous potential contribution of conservation to the energy situation. On purely technical and economic bases virtually every study completed in recent times agrees that:
- the potential for energy conservation is so great that, with only modest lifestyle changes, yet with modest economic growth, medium-term energy demand levels could be held to present levels, or perhaps even reduced, by applying the appropriate conservation measures;
- without conservation measures, no practicable policy of increased supply – renewable or conventional – can realistically and acceptably meet future energy demands; and
- investment in energy conservation measures is in general significantly more cost effective than any alternative investment in increased energy supply.

These conclusions, together with the cultural value of intelligently conserving attitudes and perceptions, lead us to stress conservation as the first measure in our alternative energy strategy.

There is no doubt that renewable energy technologies will not solve Canadian energy supply difficulties overnight. Time, substantial resources, and significant efforts on the part of many sectors of society will be required for the potential of renewable energy resources to be realized. But this does not distinguish them from conventional, non-renewable technologies and cannot legitimately be used as an argument to delay undertaking their development. The undertaking of the early nuclear energy programme was speculative, and it now emerges that perhaps twenty years or more and something like $2 billion will be required to research and develop the thorium/plutonium fuel cycle in CANDU reactor technology, let alone to finance the much larger investment in reactor technology itself. There is little reason to believe that the development of renewable energy technology will prove any more expensive over the fifty- to eighty-year life of thorium/plutonium CANDU power and many reasons to believe it will not, principally because of the relative simplicity and availability of the technology and because of the economies of scale which can reasonably be expected in manufacturing large numbers of small units such as solar collectors. In the longer term, renewable energy technology will have the great advantage of being permanently in place and capable of assimilating further refinements, while fuel constraints associated with the pursuit of the nuclear-electric option will force the need yet again to expend society's resources on the research and development of a successor technology.

This means that the question that ought to be carefully considered is whether or not the development of renewable energy resources might not meet energy requirements more readily than the development of conventional energy technology at approximately equivalent economic costs and with potentially lower environmental and social costs. Our objective here is to demonstrate that there are justifiable grounds for undertaking a concerted effort to develop renewable energy resources. In particular, we argue that there are reasonable grounds for believing that:

- Substantial government support of renewable energy is highly appropriate.
- Such support could result in a significant contribution to our national and regional energy requirements by 2025.
- Significant investment in renewable energy resources, instead of traditional non-renewable (especially nuclear) energy, is likely to be more cost effective in terms of the secondary BTU produced per dollar invested, the total amount of energy produced in 2025, the rate of expansion of energy production beyond 2025, and the substantial reduction in the environmental, economic, and social difficulties that are looming increasingly over the traditional power programme.
- Finally, when viewed in a broad, long-term perspective, there are important social and political reasons for believing that a policy that strongly supports the development of our renewable energy resources will lead to a more desirable political future for society.

Clearly, there is no single alternative energy strategy for all regions of Canada. Indeed, the strategy we discuss for Ontario is meant only to be illustrative of the range of energy options we believe to be reasonable, given the appropriate support by government and the private sector. Despite the fact that there is a range of energy futures open to Ontario, essentially only one option, the fossil-nuclear future, is being pursued. Other energy strategies we believe to be both feasible and desirable are not being seriously pursued by energy policy-makers. Since each energy strategy carries its own important implications for our environment and society, the real energy policy choices are largely social, ethical, and political, but it is precisely these dimensions that are also largely suppressed. The final choice of an energy policy should be through a well-informed democratic process.

7
Energy conservation

Canada has the second highest per capita energy consumption in the world (see Figure 1). Moreover Canadian energy consumption has been growing steadily over the years and is projected to continue to grow in the future. While much of the lifestyle we enjoy is indisputably due to the amount of energy consumed, it is becoming increasingly clear that continued growth in energy consumption will be exceedingly difficult and costly to achieve. Moreover, much of the energy produced is being wasted and therefore does not contribute to well-being. On the contrary, wasted energy is harmful. Many of the economic, social, and environmental problems that are associated with the Canadian energy dilemma could be substantially overcome by reducing the amount of energy demanded. One of the simplest and most effective ways to do this is through an increase in efficiency – i.e. a decrease in waste – with which energy is used to accomplish socioeconomic objectives.

Few would dispute that an essential goal of energy policy is to ensure adequate supply. Clearly, energy is indispensible for food, warmth, employment, and transportation. But the single goal of adequate supply is, by itself, inadequate, for it begs the question of what constitutes an adequate energy supply. There are other important social goals related to energy that must also be pursued, such as safeguarding the environment; achieving national energy self-reliance; keeping the real social and economic costs of energy as low as possible; enhancing the quality of jobs and working life; taking care that changing energy policies do not place extra hardships on the poor or unfair burdens on some regions of the nation; ensuring that future generations are not locked into bearing burdens unfairly deferred to the future; and guaranteeing that energy policy choices do not undermine the future of democracy. Historically, however, energy policy has been largely preoccupied with increasing the supply of energy available to society. The discussion in the preceding sections has illustrated how a variety of economic, social, environmental, and political constraints is now seriously challenging an energy policy that retains a preoccupation with supply. By contrast, there is a large and growing body of evidence suggesting that many of the seemingly insurmountable problems society will face if it tries simply to increase the supply of energy can be either completely avoided or substantially mitigated by reducing the demand for energy through implementation of energy conservation policies.

Many detailed studies undertaken both in Canada and the United States document the large potential for energy conservation (Table 12). Canada, like many industrial nations, unnecessarily consumes very large amounts of energy – some studies say over 50 per cent – through inefficient design of buildings and machinery. Even more energy is unnecessarily consumed by energy-intensive lifestyles. By more effectively using energy and by altering energy extravagant behaviour in modest ways, Canada could look forward to continued economic growth well into the next century without greatly increasing – perhaps even with reducing – the total amount of energy consumed.

From an economic point of view, these studies clearly indicate that by far the most cost-effective way of increasing energy resources is to stop wasting them. Dollar for dollar, effort invested in energy conservation almost always results in substantially more energy being made available to society than similar investments in the development of new energy sources. The energy resources saved by conservation will then be available for other purposes or could be left 'in the ground' for future flexibility of choice. We are thus relieved from the immediate pressure to commit enormous resources to risky and potentially damaging energy sources (Arctic fuels, nuclear electricity). In addition an effective energy conservation programme could lower the balance of payments deficits projected for the 1980s by reducing the need to import energy supplies and could eventually save consumers many hundreds of millions of dollars each year.

Conservation could substantially alleviate a number of critical factors that contribute to energy difficulties without adverse social, economic, and ecological consequences. It is important to recognize, however, that energy conservation does not by itself constitute a 'solution' to our energy dilemma. Eventually, when all areas for improving energy use have been exploited to their fullest, continued population growth and any residual per capita increases in energy that were deemed socially desirable would once again drive up the demand for energy. Conservation 'buys' time for society and frees up the necessary resources to explore and develop new sources of energy and to

TABLE 12

Selected estimates of the potential for energy conservation, various sectors and types of conservation

	Sector considered	Type of conservation	Potential savings (percentage of projected sector demand without conservation)
U.S. Office of Emergency Preparedness, Executive Office of the President, *The Potential for Energy Conservation* (October 1972)	All sectors	Elimination of waste and greater efficiency of consumption by technological and institutional means. No attempts to moderate demand as such	16 by 1980 25 by 1990
Eric Hirst and John C. Moyers, 'Efficiency of Energy Use in the United States,' *Science*, 179, no. 4080 (March 30, 1973)	a. Transportation b. Space heating c. Space cooling	a. Use of more efficient modes (no reduction in total travel) b. Increased insulation, storm windows, etc. c. More efficient air conditioners Better insulation	a. 22 to 50 when implemented b. 42 when implemented c. 67 when implemented
A.B. Makhijani and A.J. Lichtenberg, 'Energy and Well-Being,' *Environment* (June 1972)	All sectors	Elimination of waste and greater efficiency of energy use and conversion by technical and institutional means	About 68 by 2000
Charles A. Berg, 'Energy Conservation through Effective Utilization,' *Science*, 181, no. 4095 (July 13, 1973)	a. Building heating and cooling (residential & commercial) b. Industrial processes	a. Increased insulation, storm windows, control of ventilation b. Increased efficiency, heat recovery	a. 40 to 50 when implemented b. 30 when implemented

Study	Sectors	Description	Estimates

Energy Policy Project of the Ford Foundation, *Exploring Energy Choices: A Preliminary Report* (March 1974)	All sectors	a. 'Technical Fix' elimination of waste technical efficiency
		b. 'Zero Energy Growth' technical fix plus lifestyle and value changes, modified economic growth (not zero economic growth)
			a. 17 by 1985
			   36 by 2000
			b. 19 by 1985
			   46 by 2000

Policy Study Group, MIT Energy Lab. 'Energy Self-Sufficiency: An Economic Evaluation,' *Technology Review* (May 1974)	a. Industrial processes b. Cars and trucks c. Building heating (residential and commercial)	a. Eliminating heat leaks and improving waste heat recovery b. Electronic ignition, radial tires, 50 per cent reduction in automatic transmissions and air conditioning c. Thermostat adjustments, weather stripping, etc.	a. 15 to 25 by 1980 b. 20 to 30 by 1980 c. 15 immediately
Scanda Corporation Consultants to CMHC	a. Existing residences in Canada b. New residences in Canada	Insulation, ventilation changes	a. 16 to 32 depending on type of unit b. 32 to 54 depending on type of unit

F.H. Knelman, *Energy Conservation,*[*] Science Council of Canada (Ottawa, 1976)	a. Transportation in Canada
	b. Residential-commercial in Canada
	c. Industry in Canada
		a. Smaller cars, modal shifts, technical improvements
		b. Reduced heat loss or gain, improved maintenance, new techniques, design
		c. Efficiency improvements only
			a. 10 to 1985
			   10 to 1995
			b. 10 to 1985
			   15 to 1995
			c. 6 to 1985
			   9 to 1995

EMR, *Energy Conservation in Canada: Programs and Prospects* (1977)	All sectors	'Technical fix' (including attitude changes)	68 by 1990

TABLE 12 (continued)

	Sector considered	Type of conservation	Potential savings (percentage of projected sector demand without conservation)
David B. Brooks, 'A Real Option: Conservation to 1990 and Beyond' (Fall 1977)	All sectors	'Technical Fix' (economic efficiency) 'Zero or negative growth rate' (including lifestyle changes, shift to tertiary industry)	50 by 1990 ≥100 by 2025
National Academy of Sciences (Washington, 1978)	All sectors	'Technical Fix' (emphasizing efficiency, changing mix of goods and services)	66 by 2010

* Unlike other studies listed, this allows for gradual introduction of conservation measures and for their gradual acceptance; thus the results are distinctly lower than studies identifying 'potential.'

address the underlying demographic, economic, political, and social forces that drive demand for energy upward.

THE POTENTIAL FOR ENERGY CONSERVATION

Three studies of the potential for energy conservation are of particular interest: two undertaken by Energy, Mines and Resources Canada; and one done for the Science Council of Canada. Some preliminary findings of the Office of Energy Conservation* within the federal Department of Energy, Mines and Resources (OEC/EMR hereafter) are reported in *An Energy Strategy for Canada*. The role of energy conservation in residences, automobiles, and industry serves to illustrate the magnitude of the reductions in demand that can be achieved.

The greatest potential for energy savings in the residential sector lies in the more efficient use of energy for space heating. This can be attained by revising standards for new buildings, by modifying existing houses (retrofitting) and by improving the efficiency of oil furnaces, which account for over 60% of all residential heating units. Economically justified revisions in construction standards could result in energy savings in new residences of up to 50% of the energy used in similar-size residences constructed to 1970 standards. Similarly, the existing stock of houses can be improved through retrofitting to yield reductions in energy use, by 1990, on the order of 25% per unit. Oil furnaces could be improved by 20% over current efficiency levels with careful twice-a-year maintenance. Although many aspects of such programs would involve an initial capital expenditure, calculations suggest that – even at current energy prices – this increased investment would pay for itself in the form of reduced fuel costs in about five years.

[Our Figure 13] indicates the potential savings that could be achieved in 1990 through revised construction standards, retrofitting and increases in the efficiency of oil-burning furnaces. In 1975, total energy use in the residential sector is estimated at about 1 250 trillion Btu's. Of this, approximately 850 trillion Btu's (about 68%) were used for space heating. If the potential savings through energy conservation that are both technically feasible and economically justified were, in fact, to be realized by 1990, then the use of energy for residential spaceheating in that year would amount to only 70% of use in 1975, even allowing for growth in the number of housing units. Housing units built to these energy standards would have somewhat less window area, but still more than minimum standards. Otherwise, they would be no different in appearance from those typical of current

_____

* Recently renamed Conservation and Renewable Energy Branch.

FIGURE 13

Energy use for residential space heating
(Source: Reproduced from Energy, Mines and Resources Canada, *An Energy Strategy for Canada* [Ottawa, 1976], p. 93)

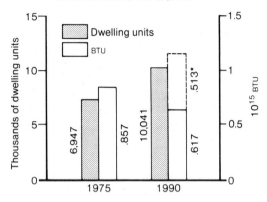

* 'Savings' attributable to improved furnace servicing (20%), new construction standards (31%), and retrofitting (49%).

construction. Major differences would be in the increased insulation of walls and ceilings, improved vapour barriers, better weather stripping, etc. Such changes could add from $200 to $1 000 to the initial cost of new housing.

Consideration of the potential for savings in the use of motor gasoline is equally striking. Recently introduced mileage standards for automobiles sold in Canada will have the effect of reducing the demand for motor gasoline in 1990 to less than 60% of what it would be with current mileage efficiency. Indeed, [our Figure 14] shows, even allowing for a normal rate of growth in the number of auto-mobiles sold, these higher efficiencies could reduce gasoline use in 1990 to a level that is about equal to gasoline sales in 1975.

In the case of industry, there is considerable potential for the reduction of energy use per unit of output, much of which will be realized in response to higher prices. Measures that can be quickly implemented and require little, if any, capital investment can yield energy savings per unit of output averaging at least 10% over a wide range of industries. Changes in energy-using processes, which would require capital expenditures and take longer to effect, could result in even more substantial reductions in energy use per unit of output. These would be par-ticularly significant for some industries and on average might amount to 15–25% per unit of output.[1]

FIGURE 14

Motor gasoline used for automobiles
(Source: Reproduced from Energy, Mines and Resources Canada, *An Energy Strategy for Canada* [Ottawa, 1976], p. 94)

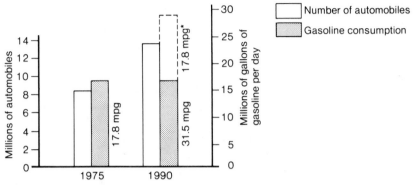

* 'Savings' attributable to increased mileage standards.

Substantial energy savings also appear to exist in the commercial sector (Figure 15). Recent energy-efficient designs of commercial buildings have an average energy efficiency of 55,000 BTU per square foot per year or less. At these levels this heat can be provided by lighting and body heat alone, without need for additional fuel. Most commercial buildings are not as efficient, around 190,000 BTU per square foot per year, and many office buildings in existence today are much less efficient at about 260,000 BTU per square foot per year. As is often the case with conservation measures, an energy efficient office building may well cost no more to build than an inefficient one, because the money saved on the smaller heating and cooling plant in the efficient building compensates for the additional costs of insulation and other energy saving technology.[2] The *Energy Strategy* concluded that, without any other conservation measures than simply the ongoing increases in the price of oil and gas (moving towards world prices), demand for primary energy would continue to grow at a rate of 3.7 to 4.3 per cent annually until 1985. These estimates are below historical rates and depend, of course, on reasonable assumptions about the overall rate of economic growth in Canada. However, in recognition of the importance of additional energy conservation measures, the document proposed a conservation target of reducing the average growth of energy use in Canada during the 1976–1986 period to about 3.5 per cent per year, representing a reduction in the

FIGURE 15
Possible impact of conservation measures on commercial energy consumption
(Source: Adapted from *Conserver Society Notes/Carnets d'épargne*, 1 [1975], p. 10.
Derived from Office of Energy Conservation, Energy, Mines and Resources,
Canada data)

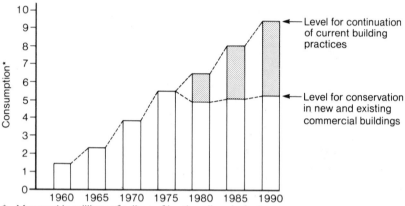

* Measured in millions of gallons of heating oil equivalent.

historical growth rate by over one-third. In order to meet this target the federal government adopted a number of specific energy conservation measures, including
− new mileage standards for automobiles;
− new guidelines for the design, construction, and operation of energy efficient buildings;
− insulation standards for residences;
− financial assistance and encouragement to improve residential insulation;
− minimum energy-efficient standards for appliances;
− an industrial assistance programme; and
− federal government 'in-house' conservation measures.
The reduction in the demand for energy during 1976 and subsequent years, presumably in large measure because of the slowdown in economic growth, and the forecasts for slow economic growth in the next decade quickly showed that the 1976 federal target of 3.5 per cent growth rate in the demand for energy would be achievable just through the gasoline savings that will result from the already mandatory increases in automobile efficiency.[3] This high-lighted that the 3.5 per cent target really did not represent a vigorous conservation effort.

Figure 7 indicates the large gap − 5 × 10^{15} BTU − that would exist between assured supplies of energy and the demand for energy associated with a 3.5

per cent growth rate. It is readily apparent that growth rates considerably lower than 3.5 per cent are required to eliminate this gap. Here is a prime example of the gulf between government statements – full of promises about potential of conservation – and the reality of present conservation policy. Modest conservation measures will have minimal impact on Canadian energy problems.

Given the inadequacy of the federal 3.5 per cent growth target, it was significant and heartening that little more than one year after the publication of the government's *Strategy* document, OEC/EMR published a further study on conservation, *Energy Conservation in Canada: Programs and Perspectives*, which at least discusses, even if it does not actively pursue, considerably more ambitious conservation targets than those of the strategy paper.[4] These targets do not constitute policy goals. Indeed, the evidence is that they are not being pursued vigorously.[5] However, consideration of this study provides insight into the policy measures that may in fact be required to address the Canadian energy dilemma.

The energy conservation measures included in this more ambitious conservation analysis were chosen to meet three criteria: technical feasibility – a conservation measure had to be either available on an off-the-shelf basis or likely to come into production during the next few years; economic feasibility – any measure in the conservation strategy had to have no more than a five-year payback time so that the investment costs associated with the conservation measure would be at least equalled by direct savings in fuel costs in five years or sooner; public political acceptability – only those measures that seemed likely to win public and political acceptability were selected for the study. Table 13 sets out these enhanced conservation measures. Beyond these, the report includes further general discussion of the potential for energy conservation through greater savings in the industrial sector by cogeneration of process heat and electricity, conservation in the energy production sector itself, particularly if costly nuclear electricity and Arctic fossil fuels are used, and in both agriculture and urban design by municipal and agricultural waste recycling, transformation to more energy-efficient technologies, and corresponding product design changes.

The important point to note is that the conservation measures examined in this OEC/EMR study were considered to be 'technically feasible and economically justifiable' today. No elaborate new technologies need be developed. No special increases in the cost of energy for the purpose of justifying these conservation measures are required. Thus, even within the narrow compass of these justifiable conservation measures, there are potentially large-scale savings to be had.

TABLE 13
Enhanced conservation measures to 1990, by sector

	Measures
Residential	Increased insulation standards in housing National retrofitting programme (70 per cent participation) Minor improvements in furnaces and appliances
Commercial	New commercial buildings to reach the standard of the existing Ontario Hydro building in Toronto Better methods of energy operation in existing buildings including better use of existing heating equipment, reduced lighting and heating at night
Transportation	Automobile fleet average mileage raised to 13 kilometres per litre Radial tires and spoilers on the cab roofs of trucks For aircraft, reduced runway waiting time and fewer first-class seats A shift away from commuting into downtown areas by private automobiles in favour of public transit
Industrial	A 25 per cent reduction in energy requirements per unit of manufactured output

Source: OEC/EMR, *Energy Conservation in Canada: Programs and Perspectives* (Ottawa, 1977).

On the basis of these enhanced conservation measures, OEC/EMR calculated an annual energy growth rate of about 2 per cent in primary energy and 1.4 per cent in secondary energy demand.[6] The important factor to recognize is that these savings were believed achievable in a situation where the population was growing at an average rate of 1.3 per cent per year to 1990 and where there was substantial real economic growth of around 4 per cent per year.

After the publication of this 1977 report on the potential of conservation, a further economic slowdown, coupled with voluntary conservation measures in response to the steadily rising, but by no means world-level, prices for energy has produced growth rates in energy demand for 1978–80 of about 2.5 per cent per year.

Again, however, the critical question is to what extent will 2 per cent growth in demand for primary energy address the Canadian energy dilemma? Reference to Figure 7 indicates that even with this lower growth rate, a large supply/demand imbalance is projected for 1990, a substantial portion of which is in the petroleum sector where upwards of 1.3 million barrels of oil each day will be required in the form of OPEC imports. Moreover, as discussed earlier, around 1990 world wide oil shortfalls are expected, and the security risks of depending on oil imports will inevitably grow, not recede, in the years ahead.

This situation underlines the importance of a much greater emphasis on energy conservation than at present and the need to establish explicit conservation goals of less than 2 per cent growth in primary energy to 1990 backed up by appropriate policy measures to stimulate economic activity. It is worth noting that the OEC/EMR study indicated that the conservation savings leading to a 2 per cent growth rate are conservative and 'should not be assumed to represent the conservation potential even to 1990. They indicate only the first cuts in energy consumption using the most obvious and most profitable measures. The full conservation potential is much higher.'[7] Note once again the gap between what government studies indicate is possible and actual policy.

The target date of both the EMR studies is 1990, which is not far away. What role may conservation play beyond that date? In 1976 a study was undertaken for the Science Council of Canada in which the effects of a variety of conservation measures were projected to 2025 in order to assess their longer term impact.[8] Two different scenarios were considered: in the first, conservation measures essentially similar to those outlined in Table 13 were continued and allowed to reach saturation – the so-called 'super technical fix' scenario (STF); in the second, additional modest adjustments to lifestyle were considered – the so-called 'conserver society' scenario (CS). The results of this study are indicated in Table 14.

In the STF scenario, even allowing for considerable growth in Canada's population, housing and automobile stocks, and industrial activity, adoption of the STF conservation measures deemed technically and economically feasible resulted in projected *total* secondary energy growth of only about 25 per cent to the year 2025. In the CS scenario, added to the features of the STF were waste recycling, the manufacture of durable goods, the widespread adoption of public transportation and some decrease in mobility and the more efficient use of commercial space, all causing a substantial absolute decrease in the secondary energy demand relative to 1973. As well, emphasis in the CS scenario is placed upon the non-energy, service sector of the economy, itself restructured to conserve energy, as opposed to the current large commitment to energy-intensive primary resource production and processing.

An OEC/EMR study team went on from their 1977 study to consider a range of energy futures beyond 1990 to the year 2025. In undertaking this exploration, the study team avoided any dramatic variations in lifestyle and assumed that Canada would remain roughly the same to 2025. However, they considered that 2025 was far enough into the future to allow for the renewal of much of society's capital in the commercial and industrial sectors (e.g., investment in housing, machines) and for most institutions to adjust to

TABLE 14
Annual energy demand, by sector, Canada (in 10^{15} BTU per year)

| | 2025 scenario* | | 1973 reference[†] |
	STF	CS	
Residential	1.08	1.08	1.09
Commercial	0.46	0.14	0.74
Industrial	2.9	1.1	1.67
Transport	1.45	0.88	1.46
Total end-use	5.9	3.2	4.96
Conversion losses	–	–	1.62
Gross primary input	5.9 + loss	3.2 + loss	6.58

* Population of 40 million. STF = Super Technical Fix; CS = Conserver Society.
† Population of 22 million.
Source: Reproduced from A. Lovins, 'Exploring Energy Efficient Futures for Canada,'
    Conserver Society Notes/Carnet D'Epargne, 1 (1976), p. 13.

TABLE 15
Indices of energy intensity in different demand scenarios (1975 = 100)

	Primary energy per capita*	GNP per unit of primary energy*
Historical		
1960	49	101
1975	100	100
1990		
Energy Strategy Report[†]	142	108
Conservation Strategy[‡] (preliminary)	113	136
2025		
High income/High industry	192	181
Medium income/Medium industry	117	209
Low income/Low industry	66	248

* For the purposes of these calculations, primary energy in the 2025 scenarios was assumed to
  equal 1.5 times the calculated figures for secondary energy.
† High price/low growth scenario.
‡ See Figure 7 and text.
Source: D. Brooks, 'A Real Option: Conservation to 1990 and Beyond,' *Alternatives*, 7
    (Fall 1977), p. 51. Figures in this table are derived in turn from the more detailed study
    by D. Brooks, R. Erdmann, and G. Winstanley, *Some Scenarios of Energy Demand in
    Canada in the Year 2025* (Ottawa: Energy, Mines and Resources Canada, 1977).

TABLE 16
Scenarios of secondary energy consumption, Canada, 2025*

	Consumption (10^{15} BTU)	Average annual growth rate, 1975–2025 (per cent)
High GNP/High industry[†]		
Residential space conditioning	0.611	
Other residential	0.859	
Commercial	1.896	
Industrial	7.144	
Automobile	0.912	
Other transportation	2.608	
Total	14.030	1.95
Medium GNP/Medium industry[‡]		
Residential space conditioning	0.486	
Other residential	0.580	
Commercial	1.086	
Industrial	3.918	
Automobile	0.674	
Other transportation	1.920	
Total	8.664	0.98
Low GNP/Low industry[§]		
Residential space conditioning	0.396	
Other residential	0.417	
Commercial	0.366	
Industrial	2.869	
Automobile	0.462	
Other transportation	1.265	
Total	5.802	−0.20

* Population of 32.181 million.
† GNP = $391.5 billion 1961 dollars.
‡ GNP = $277.2 billion 1961 dollars.
§ GNP = $183.6 billion 1961 dollars.
For comparison, 1975 GNP expanded to a population of 32.181 million = $115 billion 1961 dollars.
Source: D. Brooks, 'A Real Option: Conservation to 1990 and Beyond,' *Alternatives*, 7
(Fall 1977), p. 52. Figures in this table are derived in turn from the more detailed study
by D. Brooks, R. Erdmann, and G. Winstanley, *Some Scenarios of Energy Demand in
Canada in the Year 2025* (Ottawa: Energy, Mines and Resources Canada, 1977).

new conditions. In order to estimate future energy consumption, a range of
assumptions was made about the economic nature of Canada in the year
2025 and appropriate scenarios were generated. Tables 15 and 16 outline the
economic projections adopted and the results of the calculations. They show

a range of average annual growth rates from 1975 to 2025 of 1.95 per cent to −0.2 per cent in secondary energy consumption, depending upon the assumptions. This supports the potential long-term STF conservation magnitude estimated in the 1976 Science Council report. The amount of primary energy required in 2025 depends critically upon the efficiency with which energy is consumed, which in turn depends upon the technology used and on the energy fuels utilized, since it makes a great deal of difference in terms of primary energy requirements whether heating is derived from a solar heating system, a conventional oil or gas furnace fuelled by tar sands oil, Arctic gas, Labrador off-shore oil, or synthetic oil produced from coal, or electricity from a coal-fired generating station.

The study concluded that

over the period 1976 to 1990, Canada can easily live with 1.4% annual growth rate in secondary energy [2.1% growth in primary energy]. That is not so far above zero energy growth per capita in primary energy, a target that the Ministry of Energy, Mines and Resources has already suggested. This seems to be a perfectly feasible short-term target for Canada – provided the bureaucratic and political commitments are there ... the longer-term study shows that, over the period to 2025, achieving zero per capita growth in primary energy is really quite easy. Indeed, with moderate lifestyle changes and shift from primary towards tertiary industry, Canada could achieve zero absolute growth or even negative growth of energy consumption.[9]

ENERGY CONSERVATION AND ECONOMICS

Concern is sometimes expressed that a widespread programme of energy conservation could lead to adverse economic effects. To hold this view is to confuse energy conservation with energy curtailment. Serious economic problems will not arise from a reduction in the amount of costly energy needlessly consumed through inappropriate technological choices. On the contrary, energy conservation is more likely to stimulate an economy than to harm it. In fact, adverse economic consequences may well arise if, through failure to achieve effective conservation measures, widespread energy shortages develop and enforced curtailment of energy use results. To some extent, the failure to develop and implement appropriate energy policy in terms of price and security of supply has contributed to the climate of economic uncertainty and related difficulties facing Canada today.

Indeed, one of the most important findings emerging from the economic analysis of a number of conservation studies is that energy growth and economic growth can be uncoupled.[10] From the early 1870s to 1950, GNP per capita in the United States rose sixfold, while energy use per capita little more than doubled. Moreover, this happened at a time when the economy was rapidly shifting from agriculture to more energy-intensive industry. Economic growth far outpaced energy growth because the efficiency of energy production and use was dramatically improving. This situation has gradually changed since 1950 as overall improvements in energy efficiency have slowed down. If progress were to resume in obtaining the most useful work possible from each barrel of oil and from each kilowatt of electricity, selective economic growth could be made compatible with a resolution of energy difficulties. There is every reason to believe that a similar conclusion applies to Canada and especially to Ontario whose economy is most similar to that in the United States.

A National Academy of Sciences committee in the United States, conducting a comprehensive study of future energy options, has reached a consensus that a low rate of energy growth is possible without imposing adverse effects on the economy or requiring major changes in the lifestyle to which Americans have grown accustomed.[11] Clearly, the committee envisions the possibility of a considerably lower rate of energy growth than that suggested by most previous studies. The scenarios currently under consideration put total energy use in the United States in the year 2010 no higher than 1975 levels, and the high estimate is far lower than the figure that would prevail if historical patterns of energy growth were to continue. The committee implies that energy moderation need not entail a drastic change in lifestyle and focuses attention on increasing the efficiency of energy use and shifting the mix of goods and services towards those that require less energy. Whatever changes in lifestyle occur are expected to result from factors other than energy constraints.

These conclusions are reinforced by a report by the United States Council on Environmental Quality that claimed that the United States could increase its gross national product from 60 to 90 per cent by 2000 and at the same time reduce annual growth in energy demand to near zero. It also argued that with a maximum conservation effort the total secondary energy demand in 2010 could be reduced to 20 per cent below the 1977 level.[12] This discussion focuses attention on two dimensions of conservation policy that are generally overlooked or misunderstood.

The first dimension is that energy demand may be reduced by shifting the economy away from dependence on energy-intensive industries. The important 1974 energy policy study by the Ford Foundation considered it feasible to achieve zero energy growth after 1985, with continuing economic growth, by placing a greater emphasis on services – education, health care, day care, cultural activities, urban amenities such as parks – all of which generally require much less energy per dollar than heavy industrial activities or primary resource processing, whose growth would be de-emphasized.[13] In short, to achieve zero energy growth the United States would deliberately have to choose fewer consumer items and substitute community amenities in their place. Policies of this sort improve efficiency in the existing economy without altering it structurally. Such policies clearly require strong government leadership, but that is not beyond the scope of a modern state. This dimension of conservation policy seems, however, to be wholly ignored by both federal and provincial government statements in Canada.

A second dimension of conservation policy that is often ignored is that such a policy may lead to increased employment opportunities. Energy-intensive industry is notorious for soaking up large sums of capital to employ few people, as our present industrial technology serves to illustrate. With rising energy costs, the focus of such industrial activity will inevitably continue to shift out of the country to areas with cheaper resources and/or more readily available energy. Thus any policy capable of successfully shifting the economy to less energy-intensive activity – for example, to secondary or tertiary manufacturing – is likely both to improve international competitiveness and to increase employment. In this connection the Ford Foundation study indicated that as a result of substituting labour for energy-intensive activities, an increase in the demand for labour and potentially higher levels of employment would result.[14] Certainly, failure to increase energy efficiency is likely to reduce economic competitiveness in the long run. As Figure 1 so vividly demonstrates, successful industrial competitors who are less well endowed with energy resources have, through force of circumstance, already placed a high premium on energy efficient industry.

Such economic considerations as these are the main focus of a report by Hatsopoulos that also demonstrated a variety of potential savings through energy conservation.[15] Significantly, this study showed that in many cases it is not such economic criteria as 'expected rates of return' that constrain private investment in energy conservation, since investment in energy conservation often shows a substantially higher rate of return than does investment in increased volume of production; rather, investment decision-makers require arbitrarily that the former investments earn at a rate two times or more than investments in production before they are acceptable. Why this should be so is left to speculation, but we would attribute it to a mix of culturally-induced attitudes, perception of relative risk, effect on market presence, and tacit support for the industrial status quo. The provision of incentives for change in all of these areas will be an important part of any serious conservation policy.

This raises an issue that is examined in a careful study of the economics of energy conservation measures by Schipper and Darmstadter who argue that the current price of energy falls far below the cost of finding and developing new fossil fuel resources or of providing new large-scale electrical generating facilities.[16] For example, Ontario Hydro has interpreted its mandate – to provide power at cost – to mean that power be provided at average costs, with the consequence that the price of electricity is significantly lower than that of electricity from new nuclear plants because of the substantial contribution of low-cost hydro-electric plants which have long since paid for themselves. At present, not only do current prices tend to reflect past costs, not current marginal costs plus replacement costs, they do not take account of wider social opportunity costs, for example those associated with the future inflexibility and vulnerability of fossil/nuclear technologies. Other factors of some importance in setting inappropriately low energy prices are legal restrictions on profit rates and the structure of tax concessions available to energy corporations. Chapman, in a trenchant survey of economic disincentives to conserve energy, argues that tax writeoffs now greatly favour expansion of supply over increase in efficiency.[17] The net result of the three factors is a disincentive to conserve, a failure to relate the future costs to present consumption, and a lack of incentive to operate an economy at full efficiency.

Energy conservation policy for the United States has been carefully studied by Hannon, who focused on the interrelation between energy demand, economic performance, and employment.[18] Hannon provided a detailed list of technological shifts that would reduce energy demand and at the same time provided estimates of the jobs created (or lost) per unit of energy conserved. What emerged clearly from his analysis is the conclusion that a vigorous energy conservation policy is both possible and can be chosen so as to enhance employment levels substantially. Hannon also attached a great deal of importance to an 'energy tax' and similar economic measures to provide monetary incentives for a realistic conservation programme.

All of these contributions provide support for the reasonableness of the argument for a vigorous conservation programme based on shifts in technology and public policy, driven at least in part by economic incentives. All recognize the complexity of the problem and the necessity for action based on careful analysis, taking cognizance of the fact that values, lifestyles, and perceptions may be as important as, or more important than, technology and dollars. Government, however, does not yet appear to recognize the need, in policy and budget, to assign as high a priority to conservation as to energy supply.

IMPLICATIONS FOR ONTARIO

The projected energy demand growth rates for Canada and the United States already examined are assumed here to be directly applicable to Ontario, a not unreasonable approach considering that a third or more of the Canadian economy is situated in Ontario.

Growth curves for energy demand to the year 2025 representing the application of various possible conservation policies essentially fall between curves C and D of Figure 9. Clearly, the projected total amount of energy consumed in the year 2025 is strongly influenced by the assumed rate of growth. It is interesting to compare these demand projections with the supply of energy that was projected to be available through the 1977 expansion plans of Ontario Hydro (LRF 48A, Figure 10). Under LRF 48A generating capacity would have doubled roughly every twelve years, so that in 2008 it would represent roughly 2.0×10^{15} BTU. By 2008 growth at 3.5 per cent per annum in primary demand in Ontario gives rise to a total primary demand of 7.5×10^{15} BTU from the 1978 value of 2.6×10^{15} BTU (Figure 9), while growth rates of 2 and 0 per cent after 1977 yield demands of 5×10^{15} and 2.6×10^{15} BTU respectively. It is apparent that a lowered demand growth rate of 3.5 per cent largely undoes the effect of even the rapid nuclear expansion programme represented by LRF 48A. Only a very vigorous conservation programme, aimed at a growth rate of 2 per cent or less in primary energy demand in the context of a flourishing economy leads to a balance between demand for energy and the capacity of an energy production expansion programme to meet that demand in all-electric facilities.

This conclusion demonstrates that an essential component of any energy policy that is likely to prove socially and economically feasible is a vigorous conservation programme. There has been decided reluctance to face this fact in past discussions of energy policy, yet it is crucial to understanding the limits on practicable energy policies. The conclusion also applies, of course, to the class of energy policies based on so-called alternative energy technologies, and this is why we commence our examination of them with a study of energy conservation. However, the conclusion applies equally well to Ontario's existing policy based on conventional technologies.

In Ontario, initiative for energy conservation measures rests with the Ministry of Energy through its Energy Management Programme, which was established in March 1975 to achieve a more efficient management of energy consumption in Ontario through voluntary conservation and the encouragement of an energy conservation ethic.[19] The method employed through the programme is to calculate and demonstrate potential economies in all sectors. It is explicitly an exercise in persuasion, in the name of self-interest, confined to measures the market happens currently to make attractive. To date the budget has been modest, beginning at $1.9 million in 1975–76 and increasing to $7.2 million in 1979–80 compared with the provincial energy bill of about $7.5 billion in 1976–77, or 3,000 times the sum invested in stimulating conservation that year. The provincial energy budget is still a thousand times higher than the provincial investment in the conservation programme.

In 1979, the Ministry of Energy announced it would like to see the provincial rate of growth in demand for energy reduced to 2 per cent a year and the achievement of zero per capita growth (or about 1.2 per cent real growth allowing for population growth) in the demand for petroleum by 1985, without sacrificing economic growth.[20] The fact that, largely owing to an economic turndown, the energy growth rate in Ontario during the 1978–80 period amounted to only about 2.5 per cent, with predictions of continuing slow economic growth in the years ahead, suggests that this 2 per cent policy target, far from representing a challenging public policy goal involving a major governmental commitment, will likely be achieved through measures mostly in place at the present time and the impact of rising energy prices.

Of particular importance to Ontario is the potential for electrical energy conservation because of the central role this energy form plays in Ontario's current energy policy. Electricity is almost wholly supplied by Ontario Hydro. In fulfilling its mandate to supply electricity at cost, Ontario Hydro encouraged consumption in order to obtain significant economies of scale that translate into lower unit costs. The 'all-electric home' became a media-promoted status symbol.

Today, however, the halcyon days of cheap electricity are over. But the momentum for growth in electrical energy developed in the 'live better electrically' days is still very much present, as are the inefficient patterns of electrical energy use developed during the fifties and sixties. Ontario Hydro still operates with an outdated commercial rate structure and a base rate to municipalities proportional to average demand, both of which encourage energy consumption and discourage conservation. It is still made all-too-easy to obtain larger electrical appliances than is really necessary; advertising that favours electrical appliances, directly or by innuendo, still swamps public conservation campaigns. In view of this, a long-term reduction in electricity consumption is neither obvious to most citizens nor easy for government to bring about.

Yet reductions in demand for electricity are possible in principle and have already occurred, largely we assume because of economic adversity. The problem of efficient use of electrical energy is complicated by the fact that, unlike oil or natural gas, electricity cannot be generated at one time and then stored in large amounts for use at some other time. Rather, it must be produced at the precise moment it is needed. This is a serious problem in practice because individuals, businesses, and institutions in today's society tend to require large amounts of electrical energy at certain times of the day – the so-called 'peak periods' – and much less electricity during other – 'off-peak' – periods of the day; for example, residential consumption accounts for about 30 per cent of average demand but about 40 per cent of peak demand. Similarly there are seasonal peaks, such as in winter when heating demand is higher, that increase the peaking effect of the daily variations. The result is that some generating facilities are used at full capacity to meet the demand of those peaks for only relatively short periods each day or parts of the year. Electrical generating plants are very capital intensive and failure to utilize capacity to its fullest results in high dollar costs per kilowatt-hour being incurred. Thus the problem of more efficiently using the electrical energy system involves not only energy conservation – that is, deriving the maximum benefit from each unit of electrical energy produced – but also ensuring that the time variation in the demand for electricity is as evenly distributed as possible throughout the day – that is, reducing the demand for power during the peak periods of the demand curve.

Both in terms of its total electrical energy requirements and in the peaking problem, the electrical demand of the residential sector is of major importance. Space and water heating at present consume about 50 per cent of the electricity allocated to this sector. A further 5 per cent of the electricity consumed is for low-temperature heating or cooling applications in air conditioners (1 per cent) and for clothes dryers (4 per cent). The remaining 45 per cent of residential electrical energy is used in applications such as lighting, refrigeration, cooking, electronics, and small appliances.

It is technically feasible and, in the near future, expected to be economically attractive, to implement a variety of conservation measures for improving the efficiency of appliance and motor operation and for reduction in the levels of home lighting requirements. We might expect savings in the range of 10 to 50 per cent of current demand from these measures, depending on the appliance. Savings on cooking and lighting, which might amount to 30 per cent of current demand, would also result in substantial reductions in peak demand. Of course, a far larger impact on residential electrical demand can be made by phasing out space and water heating by electricity in favour of some renewable source. Not only could total residential demand be dropped by up to 50 per cent in this manner, but with either gas or off-peak electrical storage as back up the seasonal peak could be lowered. In sum, both average and peak residential electrical demand could be dramatically reduced if so desired. The conservation potential is there; the question remains whether there is the political, economic, and social will to realize it.

Some uncertainty exists as to the fraction of electricity consumption required for space and water heating in the commercial sector;[21] we assume a modest figure of 10 per cent, so that the scope for reducing total commercial electrical energy consumption by its replacement with renewable sources is a modest but still worthy contribution. Moreover, there is still scope for electrical savings in the remaining end-uses in this sector. Once again these fall in the range of 10 to 50 per cent of current demand, depending on the end-use. The highest demand is for lighting, generally excessive, which also makes the largest contribution to peak demand.[22]

In industry by far the largest end-use for electricity is to power motors, which amounts to about 75 per cent of the total industrial sector demand; electricity plays a relatively minor role in the production of low-temperature heat.[23] There are conservation measures to reduce lighting levels, to make the use of the electricity required in high-temperature applications more efficient, and to reduce low-level heat requirements currently being met by electricity by using either process steam or solar energy. These measures combined might result in an overall sector saving in the range of 10 to 20 per cent.

Of perhaps more substantial importance is the possibility of using co-generation technology to produce both electricity and process steam in industry to meet a significant proportion of the industrial electrical energy demand efficiently and economically. The term 'co-generation' refers to energy facilities in which fossil fuels are burned, the high temperatures thereby produced are employed to generate electricity through a variety of processes, and the waste heat necessarily rejected in this generation process is utilized to provide the process heat required by industry.[24] The use of co-generation technology significantly increases the amount of fossil fuels required to produce the same quantity of low-grade heat. On the other hand, there is an overall saving on resources for production of electricity. Heat generated in this manner could also be made available to commercial and residential sectors with appropriate urban planning. The generation of electricity in large central power plants with no heat recovery has a 'Second Law' thermodynamic efficiency of about 30 to 35 per cent. By contrast, the overall efficiency of co-generation systems is 70 to 80 per cent. Moreover, in economic terms the increased fuel costs are more than balanced off by the savings resulting from the electrical energy produced.[25]

A study undertaken for the Royal Commission on Electric Power Planning estimates that the maximum potential, if all industrial steam plants in Ontario were modified for co-generation, is about 3,000 megawatts of capacity in 1977 increasing to 4,500 megawatts by 1985.[26] The study concluded that physical, economic, and institutional constraints preclude realization of this maximum potential and suggested that about half the maximum potential might be realizable by 1985 for a total of 2,085 megawatts. This corresponds to a savings of about 50×10^{12} BTU annually (at 80 per cent load factor). If the maximum potential were realized, the industrial sector would become largely self-sufficient in electricity. The estimated realistic savings in 1985 would amount to something like 30 per cent of demand and 30 per cent of peak load projected at that time, assuming a high of 4.4 per cent growth in industrial electrical demand; with a lower growth rate, the proportional savings increase.

In the commercial sector, where steam is raised in large volumes for heating universities, hospitals, or other building complexes, the opportunity also exists for co-generation of electricity with this steam. In some situations, more electricity might be produced as a byproduct of the process than can be consumed on-site. In this situation, arrangements for marketing the byproduct power through the utility network could be arranged. For this reason, accommodations not now existing between industry and public utilities will be called for.[27]

Thus it appears technically possible and economically increasingly attractive to replace a substantial portion of the electricity consumed in the industrial and commercial sectors with co-generation technology. Studies indicate that in many applications such technology is economically very attractive, with reasonable pay-back times and total costs of the same order as investment in nuclear generating stations.[28] The co-generation process should surely be being promoted by governments with a great deal more vigour than its present 'low-profile,' low-priority treatment.

In sum, a combination of electrical energy conservation measures, the replacement of low-grade heat requirements by renewable (principally solar) energy in all sectors, and the co-generation of electricity in the industrial sector could result in overall savings in the amount of electricity consumed in the range of 25 to 50 per cent over present levels. Savings of this size should be compared with those presented in a 1977 conservation study.[29] Based on conservative assumptions, the analysis indicated that the total contribution of industrial co-generation and increased levels of insulation in electrically heated houses could equal more than 3,000 megawatts by 1989 and by 1994 the total contribution of all these measures could exceed 12,600 megawatts, or about 25 per cent of the projected total electrical capacity for that date. This is more than 1,000 megawatts more than the combined capacity of the next four large generating stations Ontario Hydro proposed to bring into service after Darlington and the study concluded that these could be postponed at this time if effective conservation policies were pursued. Though this saving falls at the low extreme of the range we have suggested as reasonable, this is attributable to the conservative assumptions made.[30] With more vigorous conservation policies assumed on the part of government and society, the savings realized would rise accordingly.

If savings of this order were accomplished 'quickly,' they would result in the following very attractive situation. Over the period to 2000, electrical demand could continue to grow at a rate of 3.5 per cent per annum and yet *all* of this demand could be met by the existing and committed electrical system. When coupled with the development of remaining hydro-electric resources and the purchase in the interim if necessary of excess hydro power from Quebec, Manitoba, and Labrador, it would be possible to allow for a substantially larger rate of growth in electrical energy demand over this same period without having to expand dependence upon non-renewable resources. Better yet, with 3.5 per cent growth in demand or less, the penetration rates of conservation practices would not have to be so high and yet expansion could still be avoided; if a higher rate of penetration were achieved to, say, 2010, then reliance on fossil fuels could be reduced in this period. Beyond this time the period could be substantially extended during which indigenous, renewable resources could be developed to meet provincial electrical demand – a topic discussed in the next chapter.

CONCLUSION

While statements about the importance of energy conservation resources figure prominently in energy policies at all levels of government in Canada, to date there exists little evidence that governments, so ready to support supply-oriented projects, are prepared to commit major resources to the conservation option. Conservation activities are receiving tiny amounts of financial support in comparison with projects directed towards increasing the supply of energy. The figures in Tables 2, 3, and 7, indicate much more clearly than any rhetoric that energy conservation is not viewed by government with the importance it deserves.

It is important to recognize that the turndown in the overall demand for energy, nationally and provincially, to a rate slightly above 2 per cent per year is not, in the main, due to the successful implementation of any major conservation program. Rather, it is mainly due to a serious slowdown in the economies of our nation and of those nations, particularly the US, with which we trade. To a considerable extent, rising energy prices and the possibility of curtailment in the supply of petroleum are contributing to the current economic malaise. There is nothing to be proud of in this reduced demand for energy, for it has been achieved at the expense of increased unemployment, rising inflation with its inequitable impact upon the poor and the old, and the foreclosing of the development aspirations of two-thirds of the world's population.

There must be no complacency. The fundamental structural changes required in industry, in the transportation sector, in the very nature of the 'consumer society' – all of which are tied into an effective and socially valid conservation programme – have not been seriously addressed by our policy makers; the constraints on national and global energy resources are not reflected in our energy and social policies. We need an energy-efficient society and the economic activities that will begin to carry us towards it. Measures must be implemented to encourage such activity, for they are necessary if Ontario and Canada are to meet successfully the challenges of the years ahead.

Even though the Conservation and Renewable Energy Branch in the Ontario Ministry of Energy has grown to become the largest section of that ministry, the amount of public investment in energy conservation remains far too small relative to the need. Resources of the same order as have been traditionally allocated for increasing the supply of energy are now required to hold back the demand for it and achieve the level of conservation necessitated by our energy predicament.

So far, the Ministry of Energy has been long on studies, short on action. The financial incentives for owners of buildings in Ontario to retrofit and so transform their energy-inefficient properties into 'conserver' buildings suitable for the year 2000 are inadequate; cars with a single occupant crowd into the major cities every morning and out of them every afternoon; energy-intensive primary industries continue to dominate while the potentially less energy-intensive and more labour-intensive secondary industries continue to falter; new housing continues to be constructed with insulation that is at best marginally sensible in terms of today's low energy costs and that will render such houses outdated white elephants in five years, consuming energy that will be a drain for decades afterwards on our energy resources. Despite its recognized value, passive solar heating still has no place in any codes or building standards; renovations of older buildings are undertaken continuously, but there is no requirement for incorporating any energy-efficient measures – another lost opportunity whose expensive consequences will last for decades.

In short, the rhetoric of conservation may impress, but the reality falls short of what is required. The potential for conservation and the ways of achieving it have been demonstrated by many detailed and extensive studies in Canada and elsewhere. The urgent need for action is underscored both by Ontario's present and long-term vulnerability and the growing gap between national supply and national demand. Public and private resources must be committed to redirect our society towards a per capita energy growth rate of near zero, but within an economy both flourishing and energy-efficient.

There is a very real danger. Canada, and Ontario in particular, stand poised to embark upon a series of extremely expensive energy projects. Conservation measures that might obviate or postpone the need for these projects must be implemented, both rapidly and on a large scale, if the growth momentum in energy demand that exists at present is to be corrected *before* serious energy shortages develop. Failure to implement effective energy conservation measures in the near future will leave any responsible politician or planner no option but to attempt to increase the availability of energy demanded by society, inevitably by embarking upon large energy projects. The costs of these supply oriented programmes, in terms of capital availability, materials, and skilled manpower, may foreclose our ability to undertake the large energy conservation programme required, thereby completing a vicious circle. They will in any case distort the economy to our detriment. Thus, if we hope to avoid the difficulties encountered with increasing the supply of energy, swift and large-scale conservation measures must be implemented immediately.

A future based upon energy conservation will require at least as much positive action by law-makers, administrators, industry leaders, and citizens as one that rests upon energy growth. To slow our growth in energy demand will require effort on a national scale. It means using energy more efficiently so that a slowdown in this sector will not seriously impair economic growth and job opportunities. Available studies convince us that energy saving is possible without either disruptive social change or unacceptable coercive government action. Indeed, in our view, the latter are more likely to be the ultimate consequence of continuing to emphasize energy supply. An effectively applied conservation policy will require skill, forethought, and consistency at every level of government, as well as ground rules that make it reasonable for energy producers and consumers to live with slower growth.

The discussion to this point has focused on the economic advantages of energy conservation. However, energy conservation is more appropriately viewed as a legitimate social goal for Canadian society rather than merely something largely resulting from other economic and energy policies. We can offer no better statement of this approach to energy conservation than that by David Brooks, former director of the Federal Office of Energy Conservation:

There are, therefore, two themes ... The first is that energy consumption is legitimately treated as a goal for Canadian society rather than as something that largely results from economic policies. The second is that the target rate of [growth in] energy consumption should be something close to zero.

These two themes can be subsumed and put into a political form by offering, as a proposition, that a lower level of per capita energy consumption in Canada is more consistent with our social well being than is a higher level, and that we Canadians can eventually achieve most of our specific social goals without significant increases in energy consumption. Putting the proposition in this way has important implications. For one thing, it implies that the rationale for energy conservation does *not* rest entirely on the adequacy or inadequacy of energy supplies. While there will obviously be differences in the severity of the problem and in the approaches to conservation depending upon supply differences, it would be nearly as important to conserve if a sudden breakthrough provided us with abundant fusion power at low cost as it will be if current pessimistic estimates of frontier oil and gas resources prove to be correct. The proposition also implies that energy conservation is not something that comes out of crisis but rather is a long-term issue. Still more, conservation involves all sources of energy and all regions of Canada. None can be exempt if the proposition is to be supported, for the implication is clearly that energy conservation is compatible with, and even supportive of, such other goals of Canadians as a higher level of environmental quality, greater attention to the demands of native peoples, and a less rapid rate of change in society. And it appears to be consistent with long-held Canadian values for maintaining diversity, small scale, and decentralization.

On the other hand, it should be made clear that the goal of moving towards zero energy growth is not intended to support a return to the low levels of energy incomes appropriate only to very simple societies, nor is it meant to ignore the fact that some Canadians still have inadequate energy available to them. Still less is the presentation of this thesis meant to ignore the very real transitional problems of moving towads zero energy growth ...

Rather, the proposition is intended to suggest that in Canada, and in other relatively rich countries, increases in per capita energy consumption have begun to
feed on themselves without providing improved services and without raising real
incomes. It is quite possible that, through the jobs-income and production-
expenditures links, we have become bemused by the sort of growth that raises the
standard of living (defined statistically as per capita gross national product) but
that simultaneously reduces the quality of life (defined loosely as opportunity for
leisure, availability of open space, privacy and so forth). If so, we appear to be
facing a difficult choice: on the one hand, the standard of living is likely to decline
(or stabilize) if growth comes to an end, but, on the other hand, quality of life
will deteriorate if growth, and particularly energy growth, does not come to an
end.[31]

This discussion focuses attention once again on what we hold central to
energy policy: the deliberate choice of a way of life in the light of both values
held and the likely ecological and economic consequences. We are less concerned with our particular preferences in energy policy, though we do not
hesitate to advocate specific policies, than we are that the people of Canada,
and those of Ontario in particular, should have the opportunity to make a
deliberate, informed choice.

8
Technology for an alternative energy strategy

Regardless of what types of energy supply technology are utilized in the years ahead, the efficient use of energy will be of central importance, for the days of low-cost energy are gone. Nevertheless, conservation by itself is not an adequate response to the energy dilemma. There remains the question of how the future demand for energy, albeit reduced in magnitude by conservation efforts, is to be met in the face of the decline in the supply of fossil fuels.

At the outset it is important to recognize that the major energy demand is not for particular energy sources as such; rather, the major energy demand is for heat – space heat, hot water, and industrial process heat – and for easily stored liquid fuels with a high energy density of which the transportation sector is the major consumer. These two kinds of energy demand account for about 90 per cent of the total (cf. Figures 2 and 16). Beyond this there is a relatively small residual set of end-uses that require the special qualities of electricity, including small electric motors, lighting, electrochemical applications, computers, and some urban transit systems.

For that matter, the ultimate social demand is not for heat, liquid fuels, and electricity either, but for the satisfactions these help to provide, such as physical temperature comfort, communication, nutrition, and interesting and convenient food. Once this is clear, it is evident that the demand for energy is not an objective quantity but rather depends on specific social value judgments. For example, the demand for liquid fuel will vary with the amount and form of transportation considered desirable. In future, the demand for liquid fuels may well be dramatically altered by changes in transportation values; this demand could be dramatically reduced by a greater use of car-pools or rail transit or substituting telephone or television linkages for business travel. Similarly, social decisions will affect every other aspect of energy demand. Though for our purposes we shall speak in terms of generic energy demands and treat their relative shares of total demand as

FIGURE 16

Canadian demand for energy by type
(Source: Adapted from R.K. Swartman, 'Alternative Power Generation Technologies,' in
Our Energy Options, Royal Commission on Electric Power Planning [Toronto,
1978], p. 77), and An Energy Strategy for Canada)

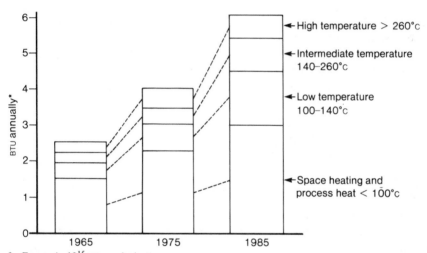

changing only slowly over our time horizon of twenty-five to fifty years, it is
important to keep in mind the social value judgments these demands express
and the fact that they are subject to change, even large change, over the
longer term.

Studies carried out both in Canada and the United States break down current patterns of energy consumption into three end-use energy categories
(Figure 16).[1] Thus our assessment of the potential of energy resources to

meet energy requirements becomes an examination of the manner in which that energy form can provide heat, liquid fuels, and 'obligatory' electricity. It is generally admitted that eventually Canada will have to rely on non-fossil sources that are either renewable or considered to be virtually limitless.[2] In this regard, there appear to be two main sources for satisfying future energy needs: nuclear-based and solar-based energy. Each of these kinds of energy source includes production of diverse energy forms for consumption so that each constitutes a potential future 'energy milieu.'

In the nuclear-based energy category are included uranium burner technology (e.g., CANDU) and thorium/plutonium, fast-breeder, and fusion reactors – the latter three all still to be developed in commercially applicable, socially acceptable forms. All of these processes can in principle supply electricity and heat, though the heat produced from electricity would be inefficiently supplied in thermodynamic terms and the waste heat from the plant would be difficult to utilize effectively. Operating on a large scale nuclear energy could even supply liquid fuels derived by electrolysis from the hydrogen in water.

It is equally important to recognize that renewable, solar-based energy encompasses various energy forms with markedly different characteristics that require different technologies for their utilization for potentially different end-uses. Energy from the sun reaches the earth as sunlight, or radiant or electromagnetic energy, that can be utilized directly in solar collectors designed to convert it into useful forms, such as heat or, through the use of photovoltaic 'solar cells,' electricity. The inflow of radiant energy from the sun also provides the driving force for the world's weather patterns. Wind energy, which can be harnessed by a variety of technologies, is one manifestation of this solar-driven weather activity. Wave energy resulting from wind action is another. The energy of falling water, harnessed in hydro-electric power stations, derives from the solar-powered hydrological cycle of the earth's atmosphere in which water is lifted from oceans and lakes as water vapour by the action of the sun's energy to be dropped as rain on upland watersheds. Radiant energy can also be captured by green plants through the process of photosynthesis, which converts it into chemical energy stored in the form of cellulose and other organic plant materials. This form of solar energy is termed biomass energy.

These various forms of solar energy are not evenly distributed over the earth's surface. Certain regions of Canada, for example, because of their latitude or relatively large number of cloud-free days, receive at ground level more radiant energy than other regions of the country. Some regions, because of their particular geographical features, are highly endowed with hydro-electric resources, experience consistently high and therefore energetic wind regimes, or have large biomass resources in the form of forests or agricultural products. It is important to note, though, that total solar-based energy is much more evenly distributed across Canada than are fossil/nuclear resources and even tends to favour the poorer, less economically developed regions; switching to solar-based energy would do much to relieve political and social stress resulting from fossil/nuclear-derived inequalities.

In Ontario, three forms of solar energy dominate: solar radiant, hydro-electric, and biomass energy. While wind power appears to have considerable potential in the northern regions of the province and in remote locations, the wind regimes of populous southern Ontario are relatively weak, and the extraction of useful amounts of energy except in certain special situations would be expensive. Wind energy, on the other hand, as well as its related energy form wave energy, appears to have considerable potential in the maritime regions of Canada.

Two other essentially inexhaustible energy forms are often termed renewable, even though they do not obtain their driving force from the sun: tidal power, whose energy derives from the gravitational energy contained within the earth-moon-sun gravitational system; and geothermal energy, the heat energy found within the earth's crust, which is believed to originate in the decay of radioactive materials deep within the earth.

We have spoken simplistically as if a nuclear-based society would use only nuclear-based energy and a solar-based society use only solar-based energy. This is in principle possible in either case, though far more practicable in the solar-based case than it is for nuclear-based energy chiefly because of the difficulty with providing liquid fuels. But probably realistic policies will employ a mix of energy resources and technologies; they can, however, be characterized as nuclear-based or solar-based depending upon which energy source dominates.

Today fossil fuels dominate. Canada, like most countries, is a fossil-based society powered mainly by oil, natural gas, and coal. Oil is the dominant energy form used in today's society, not only in terms of the total number of BTU consumed, but also because of its key importance in certain sectors such as transportation, where it provides nearly 100 per cent of the energy requirements. Oil production and distribution companies rank among the dominant economic and political institutions in the world, and institutions dependent on oil consumption such as the automobile manufacturing and service industries are major economic forces in today's society.

In a like manner, a nuclear-based industry would dominate technologically, economically, and institutionally in the nuclear-electric future contemplated for Ontario in some quarters, though of course there would at least be rather long-term reliance on some source of liquid fuel as well. Similarly, a solar era would also be characterized by a diversity of energy forms whose relative importance would shift with time though solar-based fuels and institutions would dominate. We can expect that many other energy forms would play a role in a solar energy society, some likely on a declining basis (nuclear electricity, oil, gas) and others possibly on a more permanent basis (geothermal, tidal, coal).

Such long-range energy planning as now exists strongly emphasizes nuclear energy, while according relatively little policy emphasis and research support to the solar option. While supporters of the nuclear option believe that a nuclear-electric society is both feasible and desirable, no coherent scenarios of what such a society might look like in social, environmental, and political terms have ever been presented. A crucially important question to consider here, then, is whether or not a society in which the dominant form of energy for society's requirements is solar based is both feasible and desirable.

THE SOLAR ALTERNATIVE

As is always the case when looking into the future and attempting to chart out a course of action, it is impossible to demonstrate that any one particular energy path or another will be successful. We have argued that this is certainly the case for a nuclear-electric future with its requirements for the development of new technology and its institutional and economic uncertainties. And of course it remains equally true for a solar-based energy future. What can be done is to compare systematically the nature and implications of various energy strategies open to society and then, by some evaluative process, choose among these options. In what follows we lay out the grounds for believing that in terms of technical feasibility, economic competitiveness, and potential to meet the major portion of future energy requirements, the solar option appears as reasonable as the nuclear option. Our message is that such an evaluation must be made on far more than technical or economic grounds; social, environmental, political, and ethical considerations have an equally important role to play. Indeed, if the suggestion is correct that a solar/renewable dominated future is as technically and economically accessible as the nuclear-electric future, these broader considerations should dominate the decision. We turn now to consider the solar-based alternatives.

Low-temperature heat
One of the easiest ways in which solar radiant energy can be utilized is in the production of the low-grade heat energy (temperatures of 140°C or less) that

currently available solar energy collectors can supply. Solar energy using this technology can make an important contribution to the energy needs of Ontario; just how important such contributions can be becomes apparent when it is recognized that 40 to 50 per cent of all the end-use energy produced in Ontario is in the form of low-grade heat in this temperature range. These end-uses lie in residential and commercial space heating, water heating, and air-conditioning and in industrial low-temperature process heat applications.

In order to illustrate the nature of the technology required to produce this heat energy, we examine the case of solar space and water heating in the residential sector. Two distinct approaches to solar space heating have emerged, termed 'active' and 'passive' solar heating.

One particularly attractive method for capturing solar radiation in new housing and as part of renovation activity in existing housing is to incorporate large, southern facing windows into the house design. These large windows act as collectors, allowing the solar energy to penetrate into the house, where it is absorbed as heat. In the winter, when the sun is low on the horizon, the sunlight readily penetrates deeply into appropriately designed houses, while in the summer, when the sun is higher in the sky, suitably designed roof overhangs or awnings exclude the sunlight to prevent overheating.

A variety of innovative devices are now available that help in trapping and storing heat within the house. These include special window coatings that while allowing sunlight to pass into the home, effectively block the passage of heat back out through the window. Floor and ceiling tiles have been designed that very efficiently capture and store as heat the incoming sunlight for use in the evening, while preventing possible overheating of the house. A variety of insulating shutter and curtain designs are now available that when drawn across the window at night or on heavily overcast days, prevent the loss of heat through large expanses of glass.

The advantage of this technique for gathering the sun's energy for heating is its simplicity; no rooftop collectors, large storage tanks, pumps, fans, pipes, or ducts are required. The design and construction of the house itself is responsible for the heating. This lack of mechanical apparatus has led to the term 'passive,' but it is a design policy to be as actively pursued as the 'active' approach. Passive solar heating features can be incorporated into the design of a new house for little or no additional cost, so that passive solar heating is the least expensive method for harnessing solar energy. By way of demonstrating the effectiveness of this technique, in the city of Boston a building for testing out a variety of passive solar heating technologies achieved more than 70 per cent of its heating requirements; carefully built and insulated houses in Saskatchewan achieve over 80 per cent of their annual heating requirements in this manner. Even existing buildings, however designed and heated, acquire a substantial fraction of their heat requirements from unintentional passive solar heating. A rational energy policy that could be implemented today would be to require all new and renovated buildings to incorporate passive solar energy design features. Such a policy would, relatively inexpensively, lead to immense energy savings in the future.

One device that is used in 'active' solar heating collects and converts solar radiation into thermal energy for space and water heating and is called a 'flat plate collector.' In its simplest version, this collector typically consists of a non-reflecting, black metal plate, enclosed in a glass-fronted, insulated box, that absorbs the incident solar energy and converts it to thermal energy, thereby becoming hot. This heat energy is transferred by conduction through the absorber plate to water pipes or air ducts fastened to the back of the plate and they carry the heat energy to convenient locations either for direct use or for storage and later use when the sun is not shining.

This very simple solar collector can function rather well in cases where there is a lot of sunlight and moderate temperatures. Recent engineering improvements on this basic design realize much more efficient performance and can achieve a useful output temperature even when the surrounding air is cold and considerable cloud cover is present, conditions experienced in Canada frequently. An example of this is the so-called 'tubular evacuated' solar collector in which the collector plate is located inside a transparent glass tube from which the air has been removed, much like a long thermos bottle. A solar collector of this form can gather more than 50 per cent of all the solar energy that falls upon it during the year. Temperatures of 140°C are obtainable, although temperatures of only 50°C are required for space and water heating. Active solar heating is applicable to hot water heating in both housing and commercial/industrial uses. Indeed, with the new high-temperature collectors now being developed, much of the industrial demand for heat may be met by solar energy.

A sizable heat storage system is required to provide a reservoir of surplus heat that can be used during the night and on cloudy or particularly cold days. The storage system can either be annual, where summer heat is stored in a large reservoir for winter and where little or no supplemental heating system is required, or short-term, where only a few days' reserve is maintained and for which a backup heating system is necessary to provide additional heat when required. The storage medium to which the collector heat is transferred may be water, rocks, special low-melting-temperature salts, or even wax. An ordinary forced air or hot water distribution system is used to circulate heat from the heat storage tank throughout the house, as required. At present backup systems may use natural gas, oil, electricity, or wood as fuels. In this regard we note that wood backup can be extended to urban areas a lot more effectively than is often supposed in the form of high-density compressed wood blocks. Provided electrical backup is fed directly to heat storage facilities rather than to heaters themselves, there is little or no adverse effect on the overall provincial grid requirements since the energy can then be provided off-peak and used to stabilize load demand. Such a system is already used in Australia for hot water heaters.[3] In future, conveniently stored fuels such as synthetic oil from tar sands and biomass-derived liquid fuels such as methanol and wood burned in high efficiency stoves may be appropriate backup energy sources, along with electricity.

The quality of the heat storage facilities is a key factor in the efficiency and economic competitiveness of active solar heating. We particularly wish to stress the importance of seasonal storage facilities in a climate such as Canada's. These are at present only appropriate to larger scale housing developments, such as multiple unit housing, and larger commercial and industrial developments. Even so, they would substantially decrease total backup requirements for the community. In the future it might be possible to arrange for local householder storage on a seasonal basis, by having local utilities finance the development of storage systems in each urban block. In this manner the remaining long-term backup requirements may be substantially eliminated. Over the long term such distributed seasonal storage facilities promise a system of great reliability and effectiveness with modest to low life-cycle costs. Moreover, new compact forms of heat storage are being developed which promise to reduce both the cost and the volume required for long-term storage.

Taken overall, active and passive solar collection are complementary designs, not competitors.[4] A properly designed 'passive' building reduces the overall size of the required investment in active solar collection, raises the effectiveness of the available storage facilities, and reduces the need for backup. It is important to appreciate that the 100 per cent solar heated building is realizable now. Requirements for low-grade heat in the commercial and industrial sectors could also be met using essentially these combined technologies, while air conditioning and other refrigeration needs could be met by low-grade heat-powered absorption type air-conditioning systems. Air conditioning is a rapidly growing, energy-intensive energy sector. It is therefore worth asking whether or not air conditioning is really required in well designed buildings with a climate such as Ontario's. Ontario Hydro suggests that less energy-intensive solar powered dehumidification technology may serve as a replacement.[5] Such cooling technology already exists, but at present the heat is provided by natural gas, electricity, or low-temperature process steam.

Solar heat development
The solar heating systems that have just been described are not technologies of the future. Several thousand houses throughout North America already obtain a substantial fraction of their hot water and/or space heating needs from solar energy, using similar technology. As well, a number of industries and commercial buildings are employing solar collectors. According to the Senior Adviser in the Renewable Energy Resources Branch of the Department of Energy, Mines and Resources (RERB/EMR): 'The Canadian market for low-grade (less than 100°C) heat is probably about 50% of total energy consumption ... The demand for low-grade heat, moreover, is expected to remain a roughly constant fraction of total energy consumption ... Thus the long term upper asymptote for solar market penetration potential is very high indeed.'[6]

A 1976 study undertaken for EMR examined Canada's renewable resource potential and concluded that in almost all locations throughout the most populous region of Ontario there is adequate solar energy to provide up to 100 per cent of the space and water heating needs using technology that is either commercially available today or will be in the near future.[7] In the case of new housing the installation of active or passive solar technology is relatively easy in terms of the necessary high standards of insulation, housing design, orientation relative to the sun, and on-site storage facilities. With existing housing, however, there are difficulties that limit the extent to which solar collection can be effected. Many houses are not suitably oriented, do not have appropriate locations for solar collectors, or are shaded by adjacent trees or high buildings. Informal estimates indicate that in 1980 about 50 to 70 per cent of Canadian housing stock might be suitable for solar applications with the figure rising to as high as 80 per cent by the year 2025.

These considerations suggest that the long-term potential for meeting Ontario's low-grade heat requirements from solar energy is very significant. But what are its present economic prospects? As indicated, in new housing and in some renovated existing housing, passive solar heating designs can satisfy the majority of normal space heating requirements at only a modest additional cost. The cost competitiveness of active solar technology for space heating is less clear. Much depends on the size of the building, the climate, the level of insulation, and whether or not the structure is new or is being retrofitted with solar technology. Consequently, the installed costs of systems capable of providing 30 to 60 per cent of space heating requirements currently range from about $5,000 to $12,000. The range for solar water heating systems providing 50 to 70 per cent of the energy required is about $1,500 to $3,000 (1979 dollars).

Given that any heating system, conventional or solar, is designed to provide heat over a period of about twenty years it would be misleading to make comparisons solely on the basis of initial costs. The costs of fuelling and maintaining the various systems over their lifetimes must also be considered and, of course, the reduced fuelling costs made possible by solar heating create the potential for making it cost competitive with conventional systems, the energy costs of which can be expected to escalate.

Owing to the great uncertainty over the future course of oil, gas, and electricity prices, it is difficult to prove whether solar heating is or is not already cost competitive on a life-cycle basis. The life-cycle cost of a heating system can be regarded as the sum of money that if deposited in a bank today to earn interest would be just sufficient to cover all the necessary disbursements for installation, maintenance, and fuel over the life of the system. Estimates of life-cycle costs prepared for the Department of Energy, Mines and Resources in 1978 indicated that new combined solar space and water heating systems with gas backup are already cheaper than all-electric conventional systems, assuming only a low rate of increase in the real price of all fuels.[8] Solar water heating systems appear attractive for new and some existing buildings when compared with electric systems. Since current projections of market shares see electricity as capturing some 25 per cent of the new space heating market in Ontario and more than 50 per cent in Quebec and an even greater share of the water heating markets in both, such studies indicate that the adoption of solar systems with gas backup could be a preferred option on economic grounds alone.

When it is recognized that the cost of electricity charged to the consumer is significantly lower than the cost of supplying electricity from new coal- or nuclear-fired generating stations, then the case becomes even stronger for meeting new demands for space and water heating from solar technologies using gas or even wood backup. It becomes stronger still when the ever present but largely neglected environmental and social costs of conventional energy sources are acknowledged.

Having said that in some space and water heating applications solar heating may already compete with conventional heating systems, the potential for significant gains as a result of improvements in system performance and mass production should not be forgotten. It should be noted that estimates of the cost competitiveness of solar systems versus conventional systems are very sensitive to assumptions made, including capital and installation costs. In Canada only a relatively few installations have been undertaken by any single supplier, and this lack of experience, combined with the lack of any economies of scale in the manufacture of the equipment, produces costs higher than might have been expected. Substantial efforts are already under-way in the United States, Australia, and elsewhere to overcome these pro-blems and to develop more efficient collectors and to find better storage materials. There is much less research and development of solar technology being undertaken in Canada. However, the federal government's PUSH pro-gramme committing substantial funds to the direct procurement of solar heating for federal buildings could do much to stimulate the Canadian solar market, although it remains to be seen whether the programme will be implemented on the scale announced in 1978.[9]

Finally, we emphasize again the potential for cost reductions in the long term through the installation of distributed seasonal storage facilities. Most present comparative studies do not consider such options or indeed substan-tial storage options of any sort. Since seasonal storage facilities could be expected to show low maintenance rates, long lifetimes, and to decline sharply in cost with mass deployment, they could be expected to improve life-cycle costs substantially in the long term.

A critical question that must be addressed concerns how rapidly solar energy technology can begin to be diffused throughout the economy and to contribute significantly to meeting energy requirements. We have indicated that the answer to this does not depend upon the resolution of major techni-cal problems. It depends, as we have said, on institutional commitment. One thing that has emerged clearly from several studies of the implementation of solar energy is the critical role of government in facilitating its development. More precisely, development of the solar alternative depends on the pre-sence of four factors: an appropriate legal framework, a supportive financial framework, helpful information and co-ordination institutions, and an ade-quate manufacturing, distribution, and servicing infrastructure.

The chief legal issue concerns 'rights to light' legislation. Clearly, home-owners who have installed expensive solar technology need legal guarantees that the growth of a neighbour's tree or the construction of nearby highrise buildings will not shade windows or collectors from the sun. Conversely, property holders need protection from arbitrary incursion on their develop-ment choices by the 'right to light' of neighbours. Other legal issues include taxation legislation that would encourage the adoption of life-cycle costing, development of clear and appropriate building code regulations, introduction of specific fire insurance advantages for solar-heated homes, some of which are already available, and guaranteed performance standards.

With respect to financial policy we have already pointed out the important role of adopting future-cost based energy prices and the provision of economic incentives to encourage investment in solar technologies. Possible government incentives are:
- tax deductible mortgage interest on solar systems,
- lump sum payments to adopters of solar home heating,
- tax deductibility of depreciation on solar energy equipment,
- rate subsidization of solar system mortgage loans, and
- tax deductible solar system maintenance costs.[10]

A precedent for these incentives has already been established in that the federal government and certain provincial governments have granted sales tax exemptions for solar energy equipment. On the conservation side, the Canadian Home Insulation Program (CHIP) provides very modest taxable grants for upgrading insulation in houses. At the present time the installation of solar heating technology in a private house may well lead to an increase in the assessed value of that residence, leading to higher property taxes, possibly a significant penalty for adopting solar technology. Clearly, legislation regarding property tax exemptions would help solar energy heating to compete with conventional heating systems.

Solar space heating technology involves architects and engineers, housing developers, municipal planners, the construction industry and labour unions, manufacturers, retailers, servicemen, and banks and other lending institutions. All of these people need to be well informed of the details of solar technology, the technology choices appropriate to the wide variety of needs and circumstances found, the likely financial costs and benefits involved, and the available sources for technologies and financial support. They also need a series of institutionalized working arrangements that give predictability and security to solar development as a normal part of building and renovating activity. Finally, individual property owners need similar information and security if they are to be able to make reasoned choices about solar technology.[11] At present, little or none of this institutional infrastructure exists and little is done to change the situation. It is often difficult to discover what solar technologies are commercially available, no one is coordinating the training of installation and service personnel, and there has been little or no attempt to create liaisons between financial establishments and those in the construction, planning and building fields. By contrast there is a vast bureaucracy that currently supports the development of the nuclear-electric system. In these circumstances of government inactivity, solar energy faces massive institutional barriers to its widespread adoption.

It is crucial that the manufacturing, distributing, and servicing industries develop on a sufficiently larger scale to yield both sufficient economies of scale to reduce the initial capital costs of the technology and to provide efficient installation and servicing capacity. As things stand, however, the lack of commitment from government, combined with lack of institutional infrastructure (information, education, and building codes) and appropriate financial incentives, constitutes a large hurdle to development. The economic potential is there for large-scale industrial enterprises that will help to bring employment and energy self-reliance to Canada, but the potential will not be realized unless these institutional barriers are removed.

There is a particular importance to beginning immediately the implementation of an adequate solar energy policy, since the turnover of the building stock is slow. As a senior adviser in the Renewable Energy Resources Branch of the Department of Energy Mines and Resources remarked, 'housing starts are probably at the highest point of the century now [1975] ... If substantial market penetration is delayed until 1985–90, it is clear that no great fraction of Canada's energy requirements will be met from solar in this century.'[12] A similar consideration argues for increased insulation standards effective immediately in new housing.

The policy objectives of the Ontario government with respect to the development of renewable energy in general, and of solar heat in particular, are encouraging, but the programs to achieve them are unclear. In the *Ontario Energy Review* (June 1979), Ontario is described as aiming to produce 80×10^{12} BTU per year from solar energy by the year 2000 through assisting the private sector to commercialize solar technology. From a modest contribution of around 5×10^{12} BTU in 1990, the solar component is expected to continue to rise rapidly by and beyond 2000; but the percentage share of the province's total energy demand to be met by solar heat depends on the estimates of demand in 2000. With the 'business-as-usual' scenario, solar energy would contribute 1.5 per cent, and with a scenario of zero growth in energy consumption it would contribute 3 per cent. The capital investment required from the private sector is estimated at $4.6 billion (1979 dollars). The total contribution of renewable energy, including what can be produced from wastes, to Ontario's demand in 1990 is meant to be 25×10^{12} BTU and in 2000 120×10^{12} BTU.

Three months later in September 1979, the Ministry of Energy's policy document *Energy Strategy for the Eighties*, established the new policy goal of increasing Ontario's indigenous primary production capacity by 1995 to about 60 per cent above that of the 1978 value, using provincial coal, nuclear, hydraulic and other renewable resources. Energy generated from renewable resources, including energy from wastes (municipal solid wastes, forest wastes, industrial by-product heat, and synthetic liquid fuels from agricultural wastes) plus solar, wind, and wood (both direct burning and conversion to liquid fuels or gases) is targeted to be about one-third of the overall increased self-sufficiency, the remainder being largely due to increased nuclear generating capacity. Of this renewable energy contribution, about two-thirds is attributable to solid wastes with one third attributable to solar, wind, and wood. Overall, the policy goal is to have non-hydraulic renewables meet a minimum of 5 per cent of 1995 primary energy demand in Ontario. This is a five-fold increase over the goals set out three months previously and presumably implies a major program to develop renewable resources in the province.

Unfortunately, there are no details yet announced about this program. However, the costs involved are substantial: the total effort to increase energy self-sufficiency to 1995 is nearly $30 billion which is broken down thus: $12.5 billion for the present nuclear programme, $1 billion for the lignite electrical station at Onakawana, and $16 billion for the renewable energy programme (all in 1979 dollars). This large investment could be made by individual industries, municipalities, Ontario Hydro, the provincial government, or the federal government, but the exact nature of the incentives to initiate this massive programme remain to be seen. The investment required by the nuclear programme – $12.5 billion – can doubtless be helped by government guarantees, but it is not clear what the government will do to help find the $16 billion for the renewable program or how much will have to be found by industry and home-owners. And it should be remembered that the Ministry's projection of 2,000 additional megawatts from hydraulic sources and 1,000 megawatts from Onakawana do not figure in Ontario Hydro's 1980 expansion forecast.

LIQUID FUELS

After low-grade heat energy, the second largest end-use energy requirement is for liquid fuels. The transportation sector has the greatest requirement for these fuels, amounting to 25 per cent of national energy consumption. This is a relatively inflexible demand in our economy because of the necessity to employ a fuel that can be easily stored, handled, and transported. At present, liquid fuels are generally produced from oil. Earlier we discussed the difficulties facing Canada in terms of meeting ever growing demands for oil.

Ontario, with no significant indigenous sources of oil, is especially vulnerable
and faces the serious problem of an ever growing outflow of dollars, either to
Alberta, the major source of oil in Canada, or to foreign countries. Fortunately, a potential renewable source of liquid fuels exists in Ontario in the
form of the province's biomass energy resources, which are primarily located
in the forests, though an important additional source is the large volumes of
urban and agricultural organic waste products constantly being produced in
the province.

Biomass energy
Through the process of photosynthesis, green plants convert solar radiant
energy into chemical energy and store it in the form of cellulose and other
organic plant materials. Forests, agricultural products and wastes, and urban
organic wastes (garbage and sewage) all represent biomass energy resources.
A variety of technologies already exist, or are likely to be available in the
near future, for converting biomass materials into other, more useful,
energy forms. They include:
– high-efficiency wood-burning stoves for space heating and cooking;
– generation of methane (natural gas) by 'anaerobic digestion' (bacterial
  conversion in the absence of air) of plant and animal wastes;
– large-scale combustion of wood or urban wastes to produce electricity as
  well as steam for industrial processes and district space heating (co-
  generation);
– a variety of processes which convert biomass materials to a low-energy
  content 'synthesis' gas, which can be subsequently upgraded into a variety
  of more useful fuels, including methanol (wood alcohol), a versatile liquid
  fuel.
The last is of particular relevance for the liquid fuels problem and we concentrate on it, though noting the significance of the others.
   Canadian oil production is projected to fall increasingly far below demand
over the next twenty-five years. Unfortunately, the conversion of the oil-
based transportation sector to either natural gas or electricity would be both
expensive and otherwise undesirable for a variety of reasons. Canadians also
face eventual resource shortages in the natural gas sector. In addition, engine
conversion costs and problems of safe natural gas storage in autos make
large-scale automotive conversion to natural gas a relatively unattractive
proposition. The difficulty and costs of constructing the large number of
(inevitably nuclear) electrical power plants that would be required and the
problems involved in changing over the huge auto industry to electric
motors present severe challenges to converting the transportation sector to

electrical energy. In contrast, methanol has many attractive features as a supplement to, or replacement for, oil fuels. It could be used in today's internal combustion engines with relatively minor modification. Although not without problems, the existing automobile industry and fuel distribution/ service station industry could remain largely unchanged in a conversion to methanol as a fuel.[14] Moreover, methanol produces fewer air pollutants than oil-based fuels and can lead to enhanced engine performance.

Methanol has the further important advantage that it can be derived from biomass resources that are fairly uniformly available in Canada and indeed the world. For a province like Ontario with a large demand for liquid fuels but almost no oil resources, methanol therefore represents an attractive option. The same is true or truer for the Maritime provinces.

The possibility of meeting liquid fuel requirements from biomass sources is not a matter of theoretical speculation. Brazil already produces 20 per cent of the fuel demanded in and around the industrial and commercial centre of São Paolo from ethanol (a close relative of methanol) made from sugar cane, and it is aiming to reach this proportion for the whole country within a few years. It may well be that this target will be reached sooner.[15] In the United States, is has become commercially sensible to produce 'gasohol', 90 per cent unleaded gasoline and 10 per cent ethanol made by fermentation processes from such crops as corn and sugar beet. By way of illustrating the importance of government policy in these matters, it was a federal exemption (4 cents a gallon) on excise tax and a state exemption (up to 10 cents a gallon) on sales tax that encouraged the production and consumption of this fuel with its indigenous and renewable component. The United States intends to produce $900 million gallons of gasohol in 1982 – 18 per cent of its projected demand for unleaded gasoline.

Both methanol and ethanol can be produced from plants, but methanol can be produced from other 'feedstocks' besides biomass: coal, natural gas, or even combinations such as biomass and natural gas or biomass and electricity, which can substantially increase the output of methanol per unit of biomass. The importance of the biomass and natural gas combination for Canada was noted in a study done in 1978 for the federal Ministry of Fisheries and the Environment.[16] Noting the deficiency in liquid fuel and the apparent surplus of natural gas, this study concluded that it was technically possible and economically attractive, with a price of $15 to $20 for a barrel of oil, to use natural gas and renewable biomass to produce, around the mid 1980s, up to 3 billion gallons of methanol per year; this would equal .23 × 10^{15} BTU or about 10 per cent of the projected 1985 demand for liquid fuels. Such a development would also help the forestry industry, environmental management, national security, international competitiveness, and balanced regional development, according to this study.

The study goes on to say that it would be possible for Canadian methanol production to go on expanding after 1985. In the early 1990s, there could be over 10 billion gallons produced per year, equal to $.78 \times 10^{15}$ BTU or 6.5 billion gallons of gasoline (assuming that engines run 25 per cent more efficiently when fueled with methanol); this is the equivalent of the liquid fuel from *five* tar sands plants. The possibility of this expansion was based on oil import prices hitting $25.00 per barrel (which they have already done) and on the use of trees alone without the joint use of natural gas. This study estimated that Canada's oil imports between 1985 and 1995 would exceed 700,000 barrels per day; with methanol production, this could be reduced by 110,000 to 220,000 barrels a day in 1985 and by over 550,000 barrels by 1995. After that year, forest methanol could provide such a secure energy supply that Canada could begin to export petroleum products. But, the study concluded, these things cannot begin to happen until there are clear policies, regulations, and other supportive policies established by the federal and provincial governments.

In Ontario, by far the greatest biomass resource is the forests. However, considerable uncertainty exists as to the magnitude of the annual sustainable biomass yield from Ontario's forests. Figures from the Ontario Ministry of Natural Resources indicate that the estimated annual allowable cut is about one and a half billion cubic feet. However, figures released by government agencies often account for only merchandisable pulp wood and lumber and exclude other growth that is perfectly suitable for biomass energy production. In addition, with current forest practices, a large volume of wood in the form of branches and tree tops is left in the forest. This can amount to about one-half the volume of wood actually produced but is obtainable only at extra cost.

Of course, not all this forest biomass potential will be available for methanol production. The pulp and paper industry is at present the largest consumer of forest products and the lumber industry represents an important additional demand. Wood wastes, which have historically been produced in large amounts as byproducts in forest industries, are increasingly being used as fuel in these mills and so are not likely to be available for methanol production. They can, however, contribute very significantly to the reduction of natural gas and electricity demand by this industry, which is among the largest consumers of energy in the province. In sum, forest industries will continue to place demands upon provincial forest resources.

TABLE 17
Scenarios of transportation energy demand, Canada and Ontario, 2025
(energy in units of 10^{15} BTU)

	Automobile	Other transport	Total	Ontario*
High GNP/				
High industry	0.91	2.6	3.5	1.1
Medium GNP/				
Medium industry	0.67	1.9	2.6	0.9
Low GNP/				
Low industry	0.46	1.3	1.8	0.6
Conserver society	–	–	0.9	0.3

* Calculated as one-third of the national total.
Source: Derived from D. Brooks, R. Erdmann, and G. Winstanley, *Some Scenarios of Energy Demand in Canada in the Year 2025* (Ottawa: Energy, Mines and Resources Canada, 1977), and A. Lovins, *Exploring Energy Efficient Futures for Canada*, Conserver Society Notes 1, no. 4 (1976).

However, the rate of building construction is projected to decrease and so the amount of lumber required in the construction industry may likewise decrease, though this decrease may be partially offset by increased lumber exports. In addition greater emphasis on recycling of paper products, a reduction in the use of packaging materials, and a greater emphasis on electronic data transmission and recording may result in a stabilization in the demand for paper products. And on the supply side there is growing interest in the possibility of planting and harvesting rapidly growing tree species ('energy plantations') specifically for methanol production.

The study by the federal Ministry of Fisheries and Environment estimated that in Ontario by 1985 there could be produced by the combination of natural gas and forest biomass anywhere from one to four billion gallons of methanol – i.e., about 15 to 60 per cent of the province's projected requirements for transportation fuels. Using forest products only, Ontario could also produce 2.3 billion gallons by 1985 and 3.4 billion gallons by 2005, the latter equalling .26 × 10^{15} BTU or about half of its transportation fuel requirements for 2025 under a low-growth scenario or nearly all of it in a conserver-society scenario (Table 17). The ratio between the demand for liquid fuel and the supply of biomass resources in Ontario is, however, the worst in Canada; it appears unlikely that Ontario can achieve complete self-sufficiency. However, according to this and other studies, biomass materials, chiefly from provincial forests, do have the potential to meet a substantial portion of future needs for liquid fuel in Ontario, and the remainder could be imported from other provinces in the form of methanol or oil.

In contrast to this optimistic federal study and to the Brazilian and American experience so far, a study by the Ontario Ministry of Energy, *Liquid Fuels in Ontario's Future* (1978), concluded that the province's biomass resources could replace only a small percentage of its annual oil requirements – about 6 per cent by the year 2000 and then at a cost 50 to 100 per cent higher than that of gasoline produced from crude oil at $30 (1977) a barrel. The report concluded that natural gas from the Arctic and a concerted effort to develop the tar sands would relieve the difficulties about national oil supply to such an extent that by 2000 Canada would have no need to import oil, and presumably Ontario's problem over liquid fuel would be solved. The authors of this study recommended that the Ontario government conduct no major demonstrations in the production of liquid fuel from wood, municipal solid waste, or lignite.[17]

In large measure, the differences in conclusions between the federal and provincial studies stem from different assumptions about the costs of making biomass feedstock available, the nature of the technology employed, and future costs of oil.[18] All things considered, a reasonable estimate of methanol production costs using organic wastes, including municipal and wood wastes, appear to be about twice that of gasoline on a per BTU basis. Since then, of course, the cost of gasoline has risen steeply, and it will continue to rise, which will eventually make methanol cost competitive.

The economic competitiveness of methanol production will come about through improvements in production efficiency, further increases in the price of gasoline as a result of the rising costs of oil, and/or the implementation of tax incentives that could improve the economic competitiveness of methanol production relative to that of gasoline. Price increases are virtually certain to occur, as only wholly unexpected oil discoveries could avert it; the first is also very likely, since the industry would be commencing with a virtually untried technology, and there are already indications of improvements.[19] The third alternative is of course always open to government to pursue. Moreover, there are important social and economic reasons for following through on these policies; namely, the security of supply and the employment and economic development advantages of a provincial liquid-fuel industry to the province.

A difficulty standing in the way of biomass fuel development centres on natural resources policy. Currently the forest in Ontario may have conflicting demands placed on it, as the pulp and paper, and lumber industries compete with demands for recreational areas. The demands need to be resolved in a rational resource policy. The development of a large biomass liquid fuel industry could not be undertaken without substantial rethinking by government of present resource management priorities and practices. This need for a policy review comes at a time when future uncertainties in the forest industries together with the increasing need to improve conservation and efficiency already pose major development problems. It takes considerable time for a government and civil service grown accustomed over a century to the 'simple' operation of these industries to adjust to the new realities, and this is a major hurdle for development of a biomass fuel industry.

Nonetheless a reorientation of forest resource policy may well be timely for a number of reasons. First, current forest management, including so-called 'reforestation' practices, is often unsound, leading to the destruction of a resource that ought to be renewable. Second, the pulp and paper industry in Ontario is experiencing severe economic competition from industries in the southern United States that are able to grow and harvest trees more rapidly and less expensively than can be accomplished in Ontario. The time may be appropriate to begin a government-encouraged move of the forest industry away from pulpwood and towards a biomass-liquid fuel industry with its high value-added product and assured market within the province. Hand in hand with this transition would go a much greater emphasis upon the recycling of paper and wood to reduce the demand for pulp wood and lumber. Paper can be recycled; energy cannot. Thus the arrival of the biomass fuel issue on the scene may provide the needed opportunity to achieve a new rational resource policy.

Clearly, there are a number of additional factors to be taken into account: the energy cost of obtaining the biomass; the benefit of recycling paper versus manufacturing new stock; the share of the potential biomass yield that is economically accessible and the economic impact of seasonal accessibility; the fraction of forest products required for other purposes, etc. Soil erosion, nutrient balance (especially phosphorus), concentrated truck traffic, and the ability of northern forests to produce biomass are among the environmental issues recognized as areas of concern in the development of a liquid-fuel forest industry. Even so, with appropriate conservation measures it would be reasonable to expect that a substantial proportion – perhaps as much as 30 per cent by 2000 – of Ontario's transportation energy requirements could be supplied by indigenous methanol production. With further adoption of conservation programmes, some shifts in transportation modes, modest changes in 'lifestyle,' and production efficiency improvements we might reasonably expect this contribution to double by 2025.

A study for the Ontario Ministry of Energy on the production of methanol stressed the urgency of developing the biomass alternative:

Ontario, having no indigenous liquid fuel source, a strong industrial base and a large urban region, must be regarded as a most vulnerable region in the event of a (oil) supply interruption. The industrial, commercial and ultimately social consequence of a severe shortage of transportation fuel in Ontario are unfathomable. There is therefore a compelling incentive to develop an alternative supply of liquid fuel in Ontario from indigenous biomass sources, thus reducing the Province's dependence on imported non-renewable fossil fuel resources, which in the long term must become depleted and replaced by renewable forms.[20]

Clearly, the development of a large-scale methanol industry in Ontario would reduce reliance on an increasingly costly and vulnerable supply of liquid fossil fuels and generate considerable economic activity and employment where large wood resources are located. Such regional economic development based upon indigenous resources would be highly compatible with stated economic objectives of this province.[21]

THE ROLE OF HYDRO-ELECTRIC ENERGY

The third major energy form required by society is electricity. The unique characteristics of electricity are absolutely essential for many applications; for example, lighting, motors, computers, and electro-chemical processes. However, many end-uses for electricity represent a waste of this premium form of energy; electric space and water heating are examples, as are the lighting of large buildings at night and the overlighting of stores and offices. It has been estimated that only about 10 per cent of total energy consumption falls into the category of 'obligatory' electrical energy. If electrical energy amounted to 10 per cent of total secondary energy used in Ontario in 1978, 262×10^{12} BTU would have been required. In fact, about 375×10^{12} BTU were consumed (Tables 5 and 6), an indication of the need to examine carefully the uses to which electricity is being put. The current shares of electricity by sector, and those projected by the Ontario Ministry of Energy to the year 2000, are shown in Table 18.

TABLE 18

Percentage* utilization of fossil fuels and electricity, by sector, Ontario, 1975–2000

	Gas			Oil			Electricity			Total† sector consumption
	1975	1985	2000	1975	1985	2000	1975	1985	2000	
Transportation	–	–	–	100	100 (99)	100 (98)	–	–	–	100
Residential	37	44 (44)	51 (51)	42	20 (29)	17 (16)	19	26 (26)	31 (32)	100
Commercial	47	42 (42)	42 (42)	18	15 (15)	12 (9)	35	43 (43)	46 (48)	100
Industrial	41	40 (40)	40 (39)	18	16 (16)	15 (14)	15	16 (16)	16 (18)	100

* The percentages given correspond to current trends, which include a modest conservation programme, and, in parentheses, low energy use, which includes more vigorous conservation reduction to approximately 2 per cent per annum growth in demand.
† Deviations from 100 per cent residential and transportation totals in any one year are due to an 'other' label and amount to 1 to 2 per cent; similar deviations in the industrial row are due to omission of coal, which is given as 26 per cent (1975), 28 per cent (1985), 29 per cent (2000) for the 'current trends' case, with the low energy use case differing only at 41 per cent (2000).

Source: Ontario Ministry of Energy, *Ontario Energy Review* (Toronto, 1979).

TABLE 19
Scenarios of electrical energy demand, Canada, 2025

	Total secondary energy consumption (10^{15} BTU)	10 per cent 'obligatory' electrical (10^{15} BTU)	Generation, Column 2* (megawatts)	Possible demand* (10^{15} BTU)	Generation, Column 4* (megawatts)
High GNP/ High industry (Table 16)	14.0	1.4	82,000	6.3 (45% of total = projected 2008 market share†)	370,000
Medium GNP/ Medium industry (Table 16)	8.7	0.9	51,000	2.2 (25% of total, a projected 2000 market share†)	127,000
Low GNP/ Low industry (Table 16)	5.8	0.6	29,000	0.9 (15% of total = 1980 market share)	44,000
Conserver society (Table 14)	3.2	0.3	16,000	0.3 (10% of total)	16,000

* Approximate requisite generating capacity for Column 2, respectively Column 4. Capacity for rows 1 and 2 has been first calculated at 80 per cent capacity factor and 100 per cent efficiency and then increased by 40 per cent to allow for seasonal and daily peaking and imperfect efficiency. Capacity for row 3 and row 4 has had 20 per cent added instead, because the electrical system will no longer supply low-temperature heat.

† The 45 per cent projected market share is that discussed in Chapter 4 for Ontario Hydro's generation schedule LRF 48A; the 25 per cent figure is given by the Royal Commission on Electric Power Planning as a medium-growth, medium-conservation figure; see their *A Race Against Time* (Toronto, 1978), Figure 2.6, p. 22.

As recently as 1960 almost all the electricity produced in Ontario originated from renewable hydro-electric power. Today this renewable energy form feeds about one-third of electrical demand. The remainder is provided by fossil-fuel and nuclear generating stations. We discussed the social, environmental, and economic problems Ontario faces in pursuing a path of ever greater dependence upon non-hydro-electric sources of electrical power. A question well worth considering, then, is the extent to which Ontario might be able to provide a significant proportion of its 'obligatory' electrical energy needs from its indigenous, hydro-electric resources. Table 19 indicates the amount of 'obligatory' electrical energy required in 2025 as a function of total energy demand in all of Canada. Also indicated is the amount of generating capacity required to produce that energy. The wide range in the demand for electrical energy emphasizes the critical role that assumptions about future economic activity play in influencing energy policy. By way of comparison, existing and already planned hydroelectric capacity will amount to 66,200 megawatts by 1990 – an amount in excess of 'obligatory' electricity demand in all scenarios in Table 19 except that for high GNP/high industrial growth.

Table 20 indicates the magnitude of existing and potential hydro-electric resources in Ontario. According to Ontario Hydro, only about one-half of Ontario's total theoretical hydro-electric potential is currently being harnessed. About two-thirds of the remaining potential is located in northwestern Ontario on the Albany River system.[22] But there are a number of serious problems to solve in attempting to develop these resources. Construction costs in remote northern areas are as high as or higher than those for nuclear generating plants as the James Bay hydro-electric development in Quebec has shown.[23] The transmission costs and energy losses of bringing the electricity from these northern generating sites would also be significant. There are important environmental questions that would have to be carefully assessed associated with the impoundment and flooding of the northern rivers and difficult ethical questions as to the desirability of taming one of the few remaining 'white water' rivers in the province. Finally, many of the major generating sites lie within Treaty Nine Indian Lands; the costs and benefits to the native peoples of the region, taken in the light of Quebec's James Bay hydro-electric development experience, would of necessity be the dominant social factor in undertaking the development of these rivers. Nevertheless, while recognizing the many problems, the development of some of these northern resources might well be perceived as desirable, particularly if such development were effectively used as a basis for industrial development in northern Ontario.[24]

TABLE 20
Summary of hydraulic resources, Ontario

	Average energy[†] (megawatts)
Undeveloped hydraulic sites*	
Small sites	1,369
Albany River sites	2,143
Sites unaffected by Albany River diversions	683
Sites affected by Albany River diversions	371
Total loss due to development of Albany	−216
Net undeveloped	4,350
Developed hydraulic resources	
Ontario Hydro	4,210 (6,420)
Privately developed	430 (653)
Total developed	4,640 (7,073)
Total developed and undeveloped	8,990

* Listed principally in *Water Powers of Ontario* (1946), Tables 1 and 2.
† The figures in parentheses give installed capacity. The figure of 430 MW is an estimate.

The remaining hydro sites are small relative to a nuclear power plant. Taken together, however, a number of these small generating stations could contribute significantly to regional obligatory electrical needs throughout the province. When linked, a highly secure regional grid system would be produced in which the problems associated with the failure of a very large generating plant would be avoided, including the need for backup, 'spinning' capacity required to safeguard supply in the event of the failure of a large central plant. A regional grid, while more complicated than today's relatively small number of large generating stations, is not beyond modern technological capacity and can indeed have valuable system advantages.[25] Indeed some local public utility commissions still operate hydraulic generating plants whose output is integrated with power purchased from Ontario Hydro. Also, a number of privately owned, industrial utility systems have existed for many years in the province as a result of early investment and private development of hydro-electric resources.[26] In short, there is a precedent for encouraging small, locally controlled or private electrical generating facilities in the province. Moreover, the institutional problems associated with integrating such facilities into the provincial grid appear to have been successfully dealt with.

Of particular concern, then, to this discussion is Ontario Hydro's recent trend to shut down small hydro-electric stations. The 1976 Select Committee hearing focused attention on this problem, and there are indications that certain of these small stations may be recommissioned in the future. The major barrier to the development of additional small hydro sites is the high initial capital cost. A New York study on the small-scale hydro potential in that state reached the conclusions that such developments are economically most attractive in situations where an existing, but decommissioned generating site is recommissioned, using existing facilities; less attractive when new generators are required at existing sites; and least attractive when a completely new site is being developed. If the site is developed by private interests concerned with selling electricity to customers in the immediate vicinity of the generating station, a critical factor in determining the economic viability of the project is the price the regional utility is prepared to pay to the hydro site developer for any surplus electricity available over and above that sold by the developer to his customers.[27] These problems are not insurmountable; they simply require the will to develop the appropriate solutions. Thus in the context of a deliberate shift away from large central nuclear facilities, small generating facilities may have an important future role to play. Once again the basic incentives exist: small-scale hydro-electricity is an indigenous energy source, it is renewable, secure, and can be made socially and environmentally attractive.

Conserver electrical policy
The 'obligatory' electrical component (10 per cent of total energy demand) provides a case for a long-term conserver society energy policy. If we assume Ontario's demand to be a third of the national demand, comparison of Tables 19 and 20 shows that there is at least one economically feasible course of development, – the 'conserver society' scenario – under which indigenous hydro-electric potential alone is sufficient to meet nearly all of Ontario's obligatory electrical energy needs in 2025. This is an important case for policy because of the social and environmental merits of the conserver society scenario generally and the political and economic benefits of self-sufficiency in a vital and increasingly expensive component of energy demand. However, Ontario's potential hydro-electric capacity also meets the bulk of its requirements under the low GNP/low industrial growth scenario.

It is clear that the conserver case assumes both a low economic growth rate and an extremely vigorous pursuit of electrical conservation policy; that is, other energy sources are substituted for electricity and the efficiency of the remaining end-uses is raised. Should the electrical demand increase at all significantly over the period, as in the medium-or high-growth scenarios of Table 19, even the obligatory demand would quickly outstrip provincial resources.

Assuming less than vigorous conservation, several medium-term options are open to Ontario for meeting electrical demand over and above what can be supplied by hydro-electricity and the existing and committed fossil/ nuclear generating stations, which will produce about 26,000 megawatts capacity. Of these options the most consistent would be interprovincial trade in electricity. It is important to note that in attempting to balance hydro resources with demand, Ontario presents a 'worst-case' situation, since Canada as a whole is well endowed with hydro resources. Carefully structured arrangements with neighbouring provinces could allow Ontario to receive substantial hydro-electric energy from Manitoba, Quebec, and Labrador, all regions so well endowed with hydro resources that they are currently exporting, or planning to export, electricity to the United States. Future energy policy concerned with moving towards energy self-reliance based on renewable energy resources may suggest Ontario markets for these present exports.

A second Ontario option, consistent with the first, would be electrical generation from other renewable resources – from biomass, photovoltaic cells, and in places wind. Though their short-term contribution is not likely to be large, their medium-and long-term potential is looking increasingly attractive. As for biomass-generated electricity, which already occurs in some forest industries, the Royal Commission on Electric Power Planning is sufficiently optimistic about the idea to call for a minimum of 1500 megawatts of wood-fueled electrical capacity to be developed in eastern Ontario by 2000 and to challenge Ontario Hydro and Atomic Energy of Canada to diversify their research and development expertise to meet this goal.[28]

Finally, the province could look to coal as a short- to medium-term replacement for additional nuclear capacity during the fifty-year period when its existing and committed nuclear generators will also be producing. Although coal has serious, unresolved environmental and social problems, it does have the advantage of being indigenous to Canada and in abundance. Moreover, in the United States and Europe there is a lot of research being conducted to try to control and limit the sulphur emissions which are the most serious environmental by-products of burning coal. The use of coal on a short-term basis avoids some of the intractable problems of nuclear technology. Thus coal could well be part of the 'bridge' into the medium-term future when the impact of appropriate economic policies and the development of indigenous Canadian renewable resources could be fully able to meet most of the demand for electricity.

Beyond 2000 it becomes increasingly speculative what role these options might play in provincial electrical policy. Their joint potential is very large; it is not too much to hope that, combined with an appropriate conservation policy, the province can meet all or most electrical requirements in 2025 and beyond with a combination of indigenous renewable resources and inter-provincial imports themselves based on indigenous renewable resources.

OTHER RENEWABLE ENERGY RESOURCES

There are many additional sources of solar-based renewable energy we have not discussed. We have limited detailed discussion to radiant solar, biomass, and hydro-electric energy because they are likely to make the greatest contribution in the period to 2000 and because there tend to be more technical and environmental problems connected with other sources. Additional renewable resources can only enhance our conclusion that there are many energy policy options that do not rely on fossil-nuclear energy, and that the pursuit of these should hinge on broadly based social considerations. Some of the additional energy resources available are briefly examined here.

Tidal and wave power is one option, and an initial tidal electrical generation plant is being developed in the Bay of Fundy. Wave-based electrical generation is under serious consideration in Great Britain. It has obvious potential application off both ocean shorelines of Canada and in the Great Lakes. The potential contributions from these sources are significant.[29]

There is a very large potential wind energy resource. For example, it has been estimated that there is 50 per cent more wind-generated electrical energy available on the watershed of the James Bay development (first phase) than will be generated hydro-electrically.[30] And the National Research Council estimates that full-scale deployment of windmills over those parts of Canada with reasonably high wind energy densities could supply sufficient electricity to meet all of Canada's projected electricity needs up to about the year 2000, even with rapid growth in demand. Moreover, it is not necessary to restrict wind power to generating electricity. 'If ... there is no thermodynamic advantage in making more electricity, then perhaps wind should pump heat and water and compress air rather than operate an electric grid' and in this manner contribute to a reduction of demand on other renewable energy resources. Though many commentators in the energy field hold out only a very limited role for wind-generated electricity, it is generally conceded that wind-electrical technology might have an important role to play in remoter areas. We are not sure that wind generation should be swiftly dismissed; the likelihood of a significant role for wind generated heat and electrical energy seems substantial.[31]

Photovoltaic cells, which convert sunlight directly to electricity, are a further option. Though very expensive at present, the technology is young and improving rapidly. There is substantial reason to believe that they will make a noticeable contribution to electrical power by the year 2000.[32] In the long run, the potential is there for them to supply a significant portion of electrical needs and ultimately even perhaps remove the need for electrical grids and generating stations. Though we have not considered them in our present energy policy considerations, they stand as a sharp reminder that what are now widely perceived as unavoidable energy technologies for the foreseeable future may in fact be obsolete within decades.

Heat pumps, another alternative, are used to extract and 'upgrade' to useful temperatures the readily available low-grade heat from the environment. Such pumps are more than 100 per cent efficient in that they pump more heat than they use electricity to operate. They offer important, if limited, increases in the effectiveness of heating systems in large buildings and, since they can operate when solar systems are wholly or partially inoperative, integrated solar/heat-pump systems bring closer the goal of supplying all low-grade heat requirements from indigenous sources. With local energy storage and off-peak operation of heat pumps, adverse effects on the electrical load characteristics can be avoided.

Several countries extract a substantial proportion of their energy requirements from the geothermal energy released in the earth's core at points where it is concentrated at the surface. There is a limited but real potential for similar extraction processes in Canada, particularly in British Columbia.

We have not considered such fuels as hydrogen in our discussion of future energy policies. This is because hydrogen, though a versatile fuel, is not a naturally occurring resource but must be manufactured from some primary energy form. It is not, therefore, an additional primary energy resource. At present, the most efficient way of producing hydrogen is by electrolysis, the electrical breakdown of water into its constituent elements. Large amounts of electricity are required, however; indeed, such electricity could only be provided by nuclear plants. Thus a commitment to a 'hydrogen' economy implies a commitment to nuclear power, though research is now under way to discover if hydrogen can be produced without electricity by, for example, using solar energy. Because it is easily stored and transported and could be produced from renewable solar-based energy as well as from fossil/nuclear energy, hydrogen may play an important role in developing a sustainable energy future.

THE CHOICE AMONG ENERGY TECHNOLOGIES

The preceding discussion has attempted to present some of the reasons for believing that with respect to meeting the energy requirements of an energy efficient economy in both quantity and quality, technical feasibility, and economic viability, the conservation-solar energy option is 'competitive' with the nuclear electric option. Accepting this, it becomes evident that the basis for choosing between these two options lies elsewhere: namely in terms of their environmental, social, and political implications. We shall now briefly discuss these in turn.

Environmental effects of solar-based energy
The environmental effects associated with the use of fossil fuels and nuclear energy were discussed earlier, where it was concluded that all aspects of the nuclear and fossil-fuel cycles have serious environmental problems associated with them. As our demands for energy increase, the effects associated with meeting these demands have been increasing still faster as fossil-fuel exploration is pursued in ecolocially sensitive regions such as the Arctic and offshore; as ocean oil tanker traffic has expanded with increased risk of catastrophic oil spills; and as novel forms of energy pollution (from CANDU, thorium/plutonium, fast-breeder, and fusion reactors) with hitherto unexperienced environmental problems are deployed.

In the short term, energy conservation strikes dramatically at these problems by reducing the pressures for frontier fossil fuel, for OPEC fossil fuel, and for nuclear development.[33] In the longer term, as the use of solar-based energy forms is expanded, more and more fossil and nuclear technology could be phased out, leading to the further reduction of many currently serious environmental problems.

Generally the harnessing of renewable energy resources results in a substantially reduced environmental threat compared with other energy forms. This is because, when properly utilized, the harnessing of solar energy in its many forms involves the tapping of an immense flow of energy that is already an integral part of the biosphere. Intercepting a small portion of that flow, utilizing it for some purpose, and then releasing it as low-grade heat to the environment only minimally disrupts the pattern of energy flow. This is especially true when solar radiant energy is captured for the production of heat. Compared to oil or gas heating, in a well-insulated solar-heated building a substantially reduced requirement for increasingly scarce fossil fuel exists; indeed, if the backup fuel is solar-derived methanol the fossil fuel requirement is eliminated; consequently the need for Arctic oil or gas energy systems is reduced; air pollution is correspondingly reduced; no long-lived radioactive wastes are produced by electric heating requirements.

Besides the terrestrial environment there are some important climatic consequences to current energy policies. Aerosols are produced in large quantities by the combustion of fossil fuels. Their impact on the climate is, as yet, uncertain, though some theories indicate that their effect on the atmosphere will be to exclude incoming solar radiation, thereby leading to a cooling of the earth's environment; other theories suggest that aerosols may enhance the greenhouse effect, leading to a warming effect. What is certain besides this uncertainty is that they do have a climatic impact that will increase as energy consumption increases. Conservation of energy and the use of solar energy would reduce significantly the amount of aerosol production.

Moreover, the use of solar energy avoids potential environmental problems stemming from the fact that no matter what form of energy is used, it always ends up as low-temperature heat dissipated into the environment. As noted earlier, scientists have expressed concern that the large amounts of energy now being used may alter local or regional climates and perhaps even intervene in some manner in the global climate. The use of solar energy essentially eliminates this problem.

Finally, if present energy policies are pursued there is the possibility of a change in global climate as a result of the release into the atmosphere of large amounts of carbon dioxide, an inescapable consequence of burning fossil fuels. Any such change could be extremely hazardous for human civilization, since the globe's agricultural capacity, already strained, is precariously dependent on the continuation of present climates. The use of solar energy as a major source of energy on a global scale would avoid this possibility. In certain countries of the world, large investments are currently being made in coal mining, transportation, and combustion facilities for generating electricity. Should such processes be undertaken on a global scale, large amounts of carbon dioxide would be produced. Solar energy provides a way of moving away from this situation.

The use of biomass, and especially forest biomass, as an energy source has a significant potential for harming the environment unless properly undertaken. The clearest illustration of this is the manner in which today's forest resources are being mismanaged. In effect, a process of deforestation – not reforestation – is occurring coupled with disastrous pollution of rivers and lakes by chemicals and other waste products of pulp and paper industries and lumber mills. Although much research is necessary on silviculture, especially in the demanding climate of northern Ontario, it appears possible to manage forests on a sustainable basis, practising ecologically sound planting and harvesting. In this regard, a biomass industry that can use any tree species as a feedstock, as opposed to the pulp and paper industry with its demand for soft wood, offers the potential for replanting Ontario's forests with an ecologically more desirable diverse mixture of tree species. Effective reforestation techniques, soft-wheel vehicles for harvesting and trucking biomass to central plants, and return of the mineral content of biomass to the forests are among the many aspects comprising environmentally sound policy. On the positive side, the use of biomass from municipal refuse as a feedstock for making methanol offers a way of reducing the environmental burden imposed by most forms of disposing of urban detritus.

The array of environmental effects of hydro-electric sites can be significant and include the flooding of large areas of forest or farmland, the dislocation of human settlements and animal populations, the blockage of migrating fish species, the alteration of downstream aquatic environments, and the potential for loss of life through catastrophic flooding through dam failure. Such adverse effects are particularly significant with large-scale hydro-electric projects and are clearly serious. However, dam failure is rare and by comparison with the effects of fossil fuel or nuclear energy production, most of the results of hydro-electric generation tend to be localized, predictable, and fairly short-term. In addition, the effects of any one small hydro-electric installation tend to be proportionately smaller and in many instances, through the use of proper spillway design, fish ladders or off-river reservoir-dam installations, many of these effects can be minimized. Moreover, there are also countervailing benefits in the form of reservoirs for fishing, boating, and other recreational uses and in the improved flood control achieved.

Although not without its environmental repercussions, a conserver and solar energy future can lead to a very significant reduction in harm to the environment compared with higher consumption nuclear-electric or fossil-fuel futures.

CONSERVATION-RENEWABLE ENERGY ALTERNATIVES AND
SOCIAL AND POLITICAL GOALS

The government of Ontario has issued a number of documents that set forward for consideration certain objectives relating to growth and development in the province in the decades ahead.[35] Two major themes that are intimately tied to energy policy emerge from these documents.

One of the most serious problems facing the economy of Ontario in the future is the relative decline in economic efficiency in the manufacturing sector relative to that of the province's competitors. Increased competition from elsewhere in Canada and the rest of the world will mean that the future of the manufacturing sector will depend upon its productivity. The rapid industrial development in Alberta in recent years has been markedly encouraged by the proximity of energy resources and the realization that Alberta, at least, will have energy available in the future. An associated shift of migration and economic activity to the west are a foretaste of shifting economic activity which Ontario as an energy 'have-not' province might increasingly experience in the future.

The other major problem is that of regional disparity. Ontario is a vast land mass with over 90 per cent of its population and economic activity concentrated in a small portion of its land area along the southeastern border. A wealthy urban-industrial area tends to transform the remainder of the province into an impoverished hinterland, exploited for its raw resources.[36] The reduction of regional disparity has been chosen as a major policy objective. More precisely, the government is aiming at the stimulation of economic growth in northern Ontario and in the eastern region through economic strategies based upon the resources of the area. In this regard, resource-processing activities are to be encouraged to locate near the resources rather than near the market.

It is our view that a conservation-solar energy policy fits with, and indeed reinforces, these twin economic objectives. In addition we have shown that the energy policy currently being pursued by the government of Ontario will substantially undermine these economic policies for under the present energy policy Ontario will become even more strongly dependent upon fossil fuels imported from outside its border, fuels whose cost is spiralling; upon nuclear energy whose cost is also likely to increase fairly rapidly; and upon foreign capital. The province will also develop an even more highly centralized energy system that can only exacerbate regional disparities and has an inbuilt bias towards large energy-intensive industries that undermine Ontario's competitiveness in the international market.

Now let us consider how the conserver-solar energy policy addresses these concerns. First, given that the efficiency of energy use could roughly be doubled, the cost-effective use of energy through conservation provides a major way for increasing Ontario's ability to compete internationally. Other things being equal, the higher the price of energy rises, the more effective is the policy. There will, of course, always be an economic cutoff to the economic gains available from conservation, but this limit can be expected to rise with increasing energy prices and in any case Ontario is far below the present limit.

Second, the more Ontario can base its economy upon energy resources that are independent of inevitable world price increases, the greater will be its long-term ability to compete effectively in the international market. Indigenous renewable energy forms meet this condition. Moreover, indigenous energy resources fully meet society's desire to secure material bases as independently as possible from the actions of others. This is evidently a primary motive for the development of nuclear energy in Ontario, but it applies even more to the development of renewable energy resources since uranium resources are both finite and in demand elsewhere.

Third, two of the most important renewable resources of Ontario are located in regions remote from the densely industrialized south, namely forest biomass resources and hydro-electric sites. The establishment of a methanol industry, based on comprehensive silviculture in the forested regions of the province, would provide a potentially extremely important industrial base in the northern and eastern areas. A policy of encouraging chemical industry and other manufacturing to locate there might well be undertaken together with the development of local hydro-electric and other energy sources. Potentially, a major portion of the high-grade energy needs of Ontario could be produced, on a sustained basis, in the currently disadvantaged regions of the province. The development of a large methanol industry using the biomass resources of the north and east of the province, with its assured internal market and its possibility for chemical industry, offers a way for diversifying the often narrow economic bases of our northern resource industry communities. Moreover, at the present time many of the northern regions are largely dependent upon imports of energy. This process would largely be reversed with methanol/hydro-electric development in the north. There is a real opportunity for the people of Ontario to regain a substantial measure of control over their energy and economic future through the development of a provincially owned biomass energy industry.

Fourth, because of the inherently decentralizable nature of hydro-electric, biomass, solar, and wind resources, their development tends to encourage regional development activity. This contrasts with centralized development that favours dense urban areas and is encouraged by large nuclear or fossil-fuel complexes. The former also offers the opportunity to shift economic development away from large energy- and capital-intensive industry to smaller-scale, more labour-intensive industry developed according to the needs and opportunities of regions.

Indeed, although detailed comparative data have yet to be gathered, it seems reasonable to believe, on the basis of several studies, that renewable energy technology will prove considerably more labour-intensive than has the conventional fossil and nuclear technology. In this case we can expect a considerable improvement in the employment situation for the province by moving in the direction of alternative renewable energy technologies. Two studies in the United States estimated that solar technologies will provide almost three times as many jobs per unit of energy produced as will nuclear.[37] In Ontario, the employment provided would be distributed regionally, accessible to many more people, and lack certain of the hazards associated with the centralized technologies.

Finally, we note in passing that the nature of nuclear power requires a major expansion in the civil service; indeed, about one-half of those employed in the nuclear industry are in the public sector.[38] A shift of emphasis from nuclear to renewable energy could be expected to lower this ratio substantially and with that the size of the public sector. For those concerned with burgeoning bureaucracies, increasing government spending, and the like, this may provide an additional benefit to a indigenous provincial renewable energy programme.

All the foregoing considerations apply to the other provinces of Canada, many of which have per capita richer renewable energy resources. Since it is becoming increasingly clear that pursuit of the conventional fossil/nuclear path will lead to greater national inequities of distribution and may seriously weaken confederation, the conserver/renewable path has the added advantage of supporting national unity through supporting greater interprovincial equity and autonomy.

Moreover, when Canada uses energy resources more efficiently and relies more on indigenous renewable resources it will contribute directly to easing world tensions: for example, by not using fossil fuels that are essential to developing nations; by not helping to drive up the price of those fuels, which is now costing developing nations dearly; by setting an example of careful resource management rather than one of careless greed; by not contributing indirectly to nuclear weapons proliferation through the export of its reactors, which are particularly advantageous for plutonium extraction; by not supporting international nuclear fuel reprocessing; and by not contributing to global atmospheric pollution and the dangers of climatic change. Canada's international impact is often limited in comparison to that of larger nations, but by setting an example the effect of Canadian policy may be disproportionate to the size of this country.

Beyond these restraints, we believe Canada can play an important positive world role in the development of alternative energy technology and alternative energy policy-making. Canada is a small, but industrialized resource exporting nation, heavily dependent on larger economies such as the United States, with the result that it shares some important economic structures and social interests with many developing nations. In this setting, Canada would be in a favoured position to use its own experience in the building of energy self-reliance through the efficient use of indigenous resources to aid these nations to do the same. Canada has developed an internationally recognized capability in nuclear technology, why not then in this more flexible, more environmentally and politically benign field?

CONCLUSION

Our conclusion to the chapter focuses mainly on the province of Ontario. The reader should, however, keep in mind that our intention is to suggest much broader applications.

Taken together, we believe that by 2025 the conservation renewable measures discussed would allow the province of Ontario

- to meet most electrical demand with the existing and planned hydroelectric system together with the development of additional renewable electric resources and co-generation plants;
- to meet a large fraction of liquid fuel demands from indigenous biomass resources;
- to meet a large fraction of the demand for low-grade heat from indigenous renewable resources principally radiant solar;
- to meet these energy demands in a manner that is environmentally sound, sustainable forever, supportive of regional development, employment, self-reliance, and an energy-secure and flexible future.

These measures will not resolve all of Ontario's long-term energy difficulties. There will be, for example, a residual dependence on fossil fuels to power co-generation facilities and part of the existing electrical system as well as to supply industry and to some extent the liquid fuel sector. There will likewise be a residual demand for uranium, but this would be small in relation to reserves and would eventually phase itself out as existing reactors were decommissioned and replaced by other technologies. The question of long-term energy growth has not been finally resolved; even a very small rate of growth in demand, occasioned by an increasing population or increasing use, continued long enough, will eventually spell ruin to hopes for holding the socioeconomic costs of the energy system within bounds.

While we have not provided a fully detailed description of the transitional period to 2025, we believe we have described a reasonably clear alternative energy future, achievable through the realistic development of some mix of conservation and renewable energy programmes. It is a future that in the next fifty years becomes more economically manageable rather than less so. Moreover, it is, we strongly believe, an environmentally, socially, and politically preferable future. Finally, in 2025 the province will have considerably more flexibility of choice about how to resolve the outstanding long-term difficulties than it will have if the conventional path is followed. Setting the province, and the nation, firmly on the road towards that future with the maximum number of open options is surely the least, and the most, that this generation can be called upon to do at the present time.

The challenge is to plan for our energy future: to look far enough into the future and anticipate inevitable changes in current energy trends; to understand the close interrelationship between quality of life and quality of energy form and use; to recognize our international responsibilities; to have the courage to abandon policies that have served for fifty years but now belong to the past and, even more, are a liability as we enter a new energy era; and to recognize the social implications of energy planning decisions and accordingly carefully assess the full implications of energy policies in the light of future environmental, social, and economic implications.

PART IV

FROM ENERGY POLICY TO PUBLIC POLICY

9
Institutional structure and energy policy

WHAT IS ENERGY POLICY FOR?

The usual answer to the question of what energy policy is for is that it aims to guarantee the supply of energy. We might add by way of elaboration that it aims to provide energy in sufficient quantity, at reasonable cost, and with adequate safety. This is roughly Ontario Hydro's understanding of its mandate and the view under which most other government energy agencies have worked. Thus technological and economic considerations naturally dominate the choice of energy policy.

We have argued for another view of energy policy, pointing out that there are many more social consequences flowing from the adoption of an energy policy than simply the technical provision of energy. This suggests a very different answer to the original question. We believe that the proper aim of energy policy should be the long-term enhancement of the quality of life in a society. We see energy policy as an aid to sociopolitical development rather than simply a technical response to a technical problem.

At the outset we said that energy policy stood at the heart of social policy because of the fundamental role energy plays in human life. Energy policy deeply affects all other social policy; it is a prime factor determining the quality of life members of a society experience. Some of the larger ramifications of energy policy that emerged during our subsequent discussion were the impact on

- available resources for other socioeconomic developments and the question of vulnerability to foreign control if international sources of funds and/or fuels are relied upon;
- public health and environmental quality;
- job opportunities and levels of employment;
- the potential pattern of urban and regional development;

- international environmental and political questions;
- future generations who might have to face resource and energy problems, possibly of a very large scale in a setting offering relatively little flexibility of choice; and
- the concentration of power in society and on the perceived and actual degrees of public participation.

Enough has been said about these complex but fundamental issues in the foregoing and in the literature cited in the notes and bibliography, we hope, to convince the reader that the effects of energy policy are real and at least as important to our history as those resulting from other public policies, such as those on welfare, health, and education. In the light of these considerations, we do not believe that there is any sensible course other than to agree that the overriding purpose of energy policy is to improve the quality of the lives we lead. Indeed, at this moment energy policy can in some measure allow us to mould the kind of lives we lead through the choices open to us.

INSTITUTIONS AND ENERGY POLICY

How are the fundamental decisions of which we have just spoken to be made? More pointedly, how should they be made? And once made, how do we 'get there from here'? In our view, the designs of our social institutions assume crucial importance. Social institutions form the framework within which goals and values are elaborated and tested and aspirations and perceptions of needs are developed and transformed. Within this framework, two of the indispensable roles social institutions play are the co-ordination of human activity and the resolution of human conflict. Much of the quality of life a society enjoys depends on how these are carried out. Formulating public policy is the pivotal point of the co-ordinating role of public institutions. Determining and providing for institutional means to participate in collective decision-making is the foundation for the primary, indeed essentially the only, method of humane resolution of conflict open to us. If we apply these ideas to the energy policy field there emerge four main roles for energy policy-making institutions:

- formulation of wise policy in the most appropriate terms;
- encouragement of participation and defence of autonomy in constructive conflict resolution;
- breaking negative policy reinforcements; and
- provision of transitional co-ordination and resolution of transitional stresses.

The last two functions apply to transitional periods between major policy commitments; the first two, to ongoing social activity. Though much could be said on these notions, we confine ourselves to a few brief remarks.

Wise policy

The two operative terms in our scheme are 'wise' and 'appropriate.' What is a wise energy policy is always subject to debate on the merits of the situation – as is the case for all public policies at all times. We have no general theory about wise policy and doubt that there is one. Instead we have debated the wisdom of our existing energy policy on the merits of the specific situation in which we find ourselves. However, that assessment has a wider application than just our present circumstances, since it is based on an adequate perspective, responds to the claims of all members of society in a just manner, and recognizes the subtle and complex nature of all the diverse features of energy policy. In addition, choices must not be limited in such a way that the wisest alternatives are excluded. But in the case of energy policy, we have argued that the terms in which policy is currently presented are inadequate because they are too narrow and too short-term; they neglect the larger sociopolitical and cultural issues and do not permit appropriate public participation in decision-making.

The long-term, socially oriented approach to energy policy that we advocate calls for a substantial break with past practice. Past policy documents have been dominated by economic and technological considerations in a manner that either obscured or ignored the more fundamental social considerations on which energy choices should be properly based; treated the future as a matter of prediction, in economic and technical terms, rather than as a matter of choice (under constraint) based on these social judgments; and hence were essentially 'one-way,' informing people of their future but offering no opportunity for discussion and mutual learning and choice.

We have called for the formulation of energy policy in a manner that involves abandoning each of these three features. Since the social future is not determined solely by economic and technological considerations but must be chosen, an energy policy must be decided upon as part of the deliberate realization of a certain kind of social future. Thus we need integrated, 'interdisciplinary' energy policy alternatives about which we can learn as a society through discussion and criticism.

Participation and autonomy in conflict resolution

One of the primary virtues extolled in the theory of the free-market democratic state is the promotion of individual autonomy ('consumer sovereignty'). In this setting essentially the only acceptable means of resolving conflict is independent action in the market where each person can choose where to expend his resources independently of coercion from others, and the only acceptable means of reducing given levels of conflict over scarce commodities is to raise production levels.

What this means, though, is that western industrial societies inherit two debilitating legacies: they have relatively poor institutional resources for the non-market resolution of conflict the market is incapable of resolving; and all non-market approaches to conflict resolution tend to be stigmatized as coercive and to encourage combative, adversarial strategies – and hence to be used as last-resort measures. We see the major alternative to market conflict resolution in market-dominated societies in the institutionalized adversary relations seen in the courtroom and in union-management bargaining. Just as it has long been recognized that the market cannot adequately resolve all conflicts, so it is also understood that the adversarial approach often leads to poor resolution of conflicts. We suggest that as urgently as we may need new energy resources, we need also to begin exploring new institutions that can bring richer human processes to bear on the larger public issues and can generate a cultural awareness of them and acceptance of their relevance and reliability.[1]

Breaking of negative policy reinforcements
A spectrum of energy strategies is available among which a public choice needs to be made through the open democratic process of advocation and discussion. However, it often proves extremely difficult to break out of a historical trend because of the manner in which past public policy and present culture confine present policy decisions. Governments in democracies can venture with political safety only a little distance beyond current economic constraints and mass cultural attitudes and perceptions. But both these factors are typically strongly influenced by past government and private policy. What emerges from this three-way interaction is a picture of a society bound by its own self-generated constraints, inching incrementally forward along some path chosen at each turn for its momentary attractiveness, more or less heedless of the long-term consequences until they are upon us.

Past energy policy follows such a pattern, for the origins of our energy-consumptive culture are found in past energy developments. Further, the steady incremental escalation of these policies and the culture they support will lead to severe problems now looming on the horizon. How then are we to break out of the confinements that past energy policy and present culture place on present energy decisions?

In our view the key to the problem is the nature of our energy policy-making institutions. The only acceptable way to break out of the circle and open the debate on long-term energy policy is to construct institutions that are capable of simultaneously sustaining a public and a government debate on long-term social goals and their relation to energy policy, are administratively able to draw together the entire domain into a coherent system for the purposes of this debate, and are able to administer the policies chosen. Unfortunately, Canada has little or no experience in creating this kind of institution.

Resolving transitional stress

Ontario's Energy Future envisages a turbulent transition in Ontario's energy history, extending from roughly 1980 to 2020, during which the province will make a transition from the traditional fossil-fueled, high-consumption, high waste-energy economy to some future energy economy that will supposedly be more durable with lower wastage because it is based essentially on nuclear energy resources. It is openly agreed that this transition period will be a very difficult one. Existing energy technologies to which present policy has already committed us will be competing with increasing intensity for the relatively scarce economic resources desperately needed to develop the alternative techno-economic foundation of the future. The document recognizes a real danger that society may find itself trapped between the two sets of demands, resulting in inadequate policies in both the short and long terms.[2]

What the document did not add was that there will be a corresponding competition between the 'old' and the 'new' energy institutions as they battle for control of those scarce economic resources and, less visibly but as important, the right to dominate the formulation of energy policy. Ontario stands at least as much risk of being caught between these institutional conflicts and finding itself with deeply flawed institutional structures for both purposes as the risk it runs of being caught between conflicting resource demands.

One of the key institutional roles during this transition period, then, will be to manage both the economic and institutional stresses constructively and to co-ordinate energy policy for both the short and long terms. The question that naturally arises at this point is whether our present energy institutions are suited to carry out these roles. The thrust of our critique of the policy is that they are clearly not suited to the task. We can expect the difficulties in conception and process discussed earlier to be reflected in defective institutional structural designs. By way of concrete illustration of the problems we shall briefly consider the situation in Ontario circa 1979.

ENERGY DECISION-MAKING INSTITUTIONS IN ONTARIO

Historically, provision of fuels in Canada has been in the hands of private companies, most of them foreign owned. This remains true in every field but that of electrical power; today something like 85 per cent of current energy is produced by private corporations. Each of these corporations pursues its own interests as it sees them. But while it is easy to describe these interests in vague general terms (profit, expansion, control, etc.), as soon as one attempts any description in detail it becomes necessary to engage in a detailed study of each case. Such studies are beyond the scope of this book, though we can point the reader to the rapidly growing literature in the field.[3] What is clear is that the individual interests of these companies will regularly fail to coincide with the broader public interest because they are resource exploitation institutions and society has much more complex relations to its resources than do those whose interest lies in exploiting them for profit.[4]

It is hard to assess precisely how far government can influence the behaviour of these giant companies and vice versa. They command immense wealth and employ many thousands, and they have wide international contacts and close ties with many of those in political power. They also have the advantage of a relative monopoly on expertise in their own fields.[5] Clearly, if government restricts itself to the indirect manipulation of the market, as it largely has done, then the direction of public policy will necessarily be limited to the provision of monetary incentives to these companies together with some regulatory action over them. But this certainly seems much too narrow a basis for achieving the kind of broad policy objectives we have argued are essential for energy policy. A sound institutional proposal for future energy policy-making institutions must confront directly the relationship between public and private energy policy and the resolution of conflicts between the two. We believe this can be accomplished in a constructive manner.[6]

Ontario Hydro

Another energy institution is Ontario Hydro, which has wide authority in relation to the generation, transmission, and distribution of electric power and indeed all forms of municipally managed energy throughout the province of Ontario. Hydro considers its role to be 'the provision of generation and transmission facilities to supply the electrical power demands of the people of the province.'[7]

However, Hydro's mandate cannot be interpreted in this narrow commodity-oriented manner. A 1972 government report on Ontario Hydro specifies the objectives of the corporation as follows: to meet demand at lowest feasible cost; to fulfil demand reliably; to support regional development policies; to stabilize business cycles; to support environmental policies; to exploit new technologies; and to stabilize capital markets. Such an encompassing list of goals is already difficult to reconcile with Hydro's own view of its mandate and with the frequent admonitions, from both government and Hydro officials, to keep government policy from interfering with the corporation's market activities.[8]

Ontario Hydro is institutionally structured around the design, construction, and maintenance of its generation and transmission technology. There is no doubt that Hydro has a technical record equal to that of any utility in the world. However, policy is determined for the entire institution, and hence for the province, by a small executive group at the head of the institution. Below this level, official discussion about policy issues, whether within the institution or with government agencies and municipalities, is minimal.[9] In short, a familiar picture emerges of the market institution, which is centralized, technically oriented, and efficient in the production of its commodity but structurally unable to relate to the wider social ramifications of its activity.

Like any private company Ontario Hydro tries to predict future market demand. This approach stems from assumed 'consumer sovereignty' in the market. Such an assumption is at least partially illusory and becomes increasingly so as the social importance of the 'commodity' in question increases. Recognition of this fundamental circumstance has come slowly but is increasingly reflected in government approaches to policy. Nevertheless, there is still an enormous gap between the imperative of the market – namely to plan, to produce, and to meet expected market demand – and the much wider consideration of the future as a matter of deliberate choice in the public domain. The latter objective can be achieved in two radically different ways. The way traditionally followed in electric power planning has always been to make future *consumption* a matter of choice by over-supplying. The alternative is to make future *production* a matter of choice by flexible planning, production, and energy technology strategies.

Efficiency, growth, and control – the dominant goals of market-oriented institutions – lead to centralized, high-technology institutions with dominant experts and passive clients. Hydro is no exception, for the thrust of research and development is towards ever larger, ever more technically sophisticated, centralized production plants. Hydro has been actively phasing out small generation plants; there is no planned role for local generating plants based on biological energy or any involvement in solar technology – in contrast to US utilities which are active in making energy conservation advantages and solar technology available to their customers. Hydro expansion plans call instead for large nuclear power plants with all of their attendant consequences.[10] And it is characteristic of Ontario Hydro that, in a publication on long-range power planning, solar energy is mentioned only in passing and then only to refer to electric solar cells.[11] Instead, the Hydro publication discusses the consumer virtues of electric heating, omitting consideration of wider social costs. In partial anticipation of arguments that other alternatives are socially preferable, Hydro claims the moral immunity of the market by invoking the sovereignty of consumers as 'public demand' and 'industrial need.' What this amounts to in reality, however, is the acceptance of consumer preferences that Hydro itself has helped to create.[12]

The associated removal of decision-making control from the public domain is well illustrated by the recent efforts of Hydro to explain its evaluation of the social consequences of future power developments.[13] The most crucial social issues are simply ignored, and all but strictly financial and technical issues are rather summarily dismissed. The socially important decisions are to be matters for internal judgment: 'While economic differences can be quantified [*sic*!] many other differences cannot. As a result the trade off process requires considerable judgement. *Control of this judgement is obtained by corporate review of the proposals* before major alternatives are committed for design and construction.'[14] That judgment clearly has a bias towards 'quantifiable' economic factors. Unfortunately, we are not told what the process of corporate judgment entails. Nevertheless, we would rather have internal decision-making replaced by public choice, institutionalized in a public, critical process. But Hydro's approach to public participation, as revealed in the document *Public Participation*, does not leave much room for a consultative process: 'Often, public participation has been against the result of the study (a site or line location) rather than the study process. This suggests that a more realistic understanding of the role of public participation is required *so that the process of involving the public is not considered a replacement for decision making.*'[15]

Ontario Hydro is only continuing the tradition, widespread in the Canadian nuclear field, of bypassing public involvement. Only recently has Hydro introduced any public hearings on siting of nuclear installations and these have come after all the major issues have been privately settled. The public has never been involved in any decision of Atomic Energy of Canada Limited or, to any appreciable extent, in the Atomic Energy Control Board's hearings and decisions about the design, siting, and licensing of nuclear installations. The Interim Report of the Royal Commission on Electric Power Planning called for such involvement.[16]

Other efforts of the government to involve the public through hearings have as often been directed mainly at informing itself of the plans of the giant energy bureaucracies. Technical reports have simply been presented to a passive public. Indeed, there could be no better illustration of the ad hoc nature of critical inquiry in a centralized institution than the recent parade of board hearings, committees, and commissions, each focusing on some narrow aspect of energy policy and each incapable of formulating coherent energy policy.

Ontario Ministry of Energy
The Ministry of Energy was established in 1973 with the specific objective to

make recommendations for the effective coordination of all energy matters within the Government of Ontario with a view to ensuring the consistent application of policy in every area of concern regarding energy ... to make recommendations regarding priorities for and the development of research in all aspects of energy of significance to Ontario, including the conservation of energy and the improvement of efficiency in its production and utilisation and the development of new energy sources.[17]

These, in addition to the general objective to 'review energy matters on a continuing basis with regard to both short-term and long-term goals in relation to the energy needs of the Province of Ontario,' are wide ranging responsibilities, even if the letter of the law is vague about the way energy 'needs' are to be understood or determined (after all a matter of crucial importance!).

But the conspicuous lack of political and regulatory power given to the Ministry at present makes the wide mandate rather unconvincing. The Ministry's role is limited to providing advice and assistance to the government, which are hardly encompassing powers for a ministry in charge of regulating Ontario Hydro (through the Power Commission Act and the Ontario Energy Board Act). As wide as the Ministry's terms of reference seem, its powers are narrow. This places the Ministry of Energy in quite a different position from, for example, the Ministry of Health which can wield far more political power.

In practice, too, the policy of the Ministry of Energy seems to be strongly oriented towards fossil fuels, complementing the effective monopoly Ontario Hydro has over the field of nuclear-generated electricity. The overall orientation of the Ministry to the supply of energy was well illustrated in its policy document, *Ontario's Energy Future*,[18] though most of the Ministry's budget is now spent by its Conservation and Renewable Energy Branch. Furthermore, real political control over energy supply and development in the province, in so far as it exists, is evidently exercised by other ministries or government institutions, only one of which – the Ontario Energy Board – falls under the Ministry's authority.

The Ministry of Treasury, Economics and Inter-governmental Affairs (TEIGA)
TEIGA has many of the important means to develop and administer an
effective and integrated energy policy through its powers over regional development and infrastructure, finances, management of the provincial debt,
contact with local governments, and establishment of regional governments.
The orientation of the Ministry, however, is heavily biased towards financial
and economic matters, as a glance at the Ministry's internal structure will
show.

The management of the provincial debt is one of the responsibilities of the
Treasurer, and since most of the province's borrowing is on behalf of
Ontario Hydro, TEIGA is in a strong position to control the extent, if not the
form, of Ontario Hydro's spending. As the Ministry recognizes, 'the planning and implementation of Ontario Hydro facilities could play an important
role in supporting and shaping the economic development of Ontario ... The
role of Treasury with respect to these issues is to provide both broad policy
directions and advice on specific issues primarily through the Advisory Committee on Regional Development.'[19]

In view of this it is rather surprising to read in the same document that the
Ministry's contact with Ontario Hydro consists of the Treasurer meeting
with the Chairman of Ontario Hydro to 'ensure that the structure and timing
of Ontario Hydro's financing is consistent with the overall fiscal policy of the
Province.'[20] Although there are other, more indirect contacts between the
two institutions, we are dismayed by the inconsistency between the stated
aims of the Ministry and the practice of its relations with Ontario Hydro.
However, this situation would be understandable if we were to assume that
energy were simply a marketable commodity and Ontario Hydro an optimal
allocator in the light of market forces. It would also be understandable if
Ontario Hydro were relatively impervious to attempts at effective government control.[21]

Cabinet Committee on Resource Development (CCRD)
TEIGA and the Ministry of Energy maintain contact through CCRD, whose
three specific functions are stated to be to carry out independent evaluations
of proposals from member ministries, to create a liaison among ministries,
and to develop policy that encompasses more than one ministry. Especially
the latter function should bring CCRD squarely within the realm of energy
planning; yet the Committee is in an unfortunate position with regard to
energy policy development as well as implementation, since its structure and
place within the government allow it only to advise and recommend. In the
apparent absence of any government institution clearly charged with energy

policy implementation we are left to wonder who the recipient of the numerous recommendations is supposed to be.

CCRD is backed by the Resources Development Secretariat (RDS) to co-ordinate resource development policy. At this level there seems to be some opportunity to achieve wider policy integration. But this is still limited to the resources fields, where all natural resources are lumped together. Moreover, the Secretariat, too, is limited to an advisory role, albeit a potentially important one, given the opportunity to influence policy development.

Perhaps a co-ordinating organ at the level of CCRD/RDS represents an appropriate form of interministerial contact. As far as we can see, however, RDS and CCRD are unconnected to appropriately designed policy-making institutions with enacting power, and the current channels of communication are not appropriately related to the real responsibilities of the resource policies involved.

Ontario Energy Board (OEB)
The OEB reviews electricity prices and has a regulatory role over natural gas price and oil and gas transmission lines. Since the Board's criteria for allowable rate increases recognizes the effect of costs prudently incurred, there is some scope for policy directives with regard to financial matters. To our knowledge this opportunity has not yet been used to any appreciable extent, as the Board has explicitly confined itself to the economics of rate increases based upon predicted future demand and costs.[22]

Ministry of the Environment (MOE)
MOE has some measure of regulatory control over Ontario Hydro in matters pertaining to environmental standards. Specifically, the Environmental Assessment Act gives the Ministry power to obtain and evaluate a statement of socio-environmental impact assessment for each of Ontario Hydro's capital installations. Whether this power is consistently exercised remains an open question, since its use depends upon the discretion of the Minister and the next large nuclear installation, Darlington generating station, was exempted from the demands of the Act by ministerial decree in 1977.

From the preceding discussion it is clear that such powers are, even so, far too narrow in scope and of the wrong kind to constitute an effective integrated policy-making tool. One can fairly conclude, therefore, that as important as these assessment procedures are, to use the existing institutional and legal framework for the implementation of overall energy policy goals would be neither efficient nor effective.[23]

Apart from the government institutions there have been several other bodies in the field of a more temporary and ad hoc nature, such as the Solandt Commission, the Advisory Committee on Energy, Task Force Hydro, the Select Committee of the Legislature Investigating Ontario Hydro, and the Royal Commission on Electric Power Planning. None of them is or was in charge of making or implementing policy. Their terms of reference allow(ed) them to influence the political climate and to prepare the scene for future policy changes – and many of them have indeed done so – but it is up to the more permanent sections of government to take policy decisions. Nevertheless, the comments and the recommendations that have come forth from the various committees and commissions have not been without practical effects, although institutional structures have to a large extent remained as they were, despite some rather serious criticism.

At the regional and local level the relevant non-government institutions are private industries and local utility commissions. In our rapidly centralizing world, however, strictly local private industry is becoming somewhat of an anachronism; rather, large, centrally directed companies predominate. Municipal energy utilities are obviously restricted to a quite narrow range of locally relevant policy options. In fact, they tend to be little more than centrally directed energy retailers and, because of an energy generation system that is becoming increasingly centralized, they are seldom in a position to influence energy planning significantly.

Municipal governments in Ontario are in an unenviable position. Local governments lack the economic base to do much more than maintain roads and collect garbage. As such they tend to be glorified utilities, their status as 'creatures of the province' enshrined legally and their responsibilities often partial and ambivalent. This situation is also enshrined economically, since their revenues are restricted to a property tax base. And important social functions, such as the control of education, health and welfare, and energy, are in some degree institutionally independent of municipal governments.

Failure of Ontario institutions

What emerges from this brief review is a picture of a welter of institutions, public and private, transitory and permanent, each in some interaction with some of the others, a web without a coherent policy-making focus. In fact we have no better words in which to state our general conclusion than those of the Select Committee of the Legislature Investigating Ontario Hydro:

The Committee found, however, that while the public – directly and through its Government – was involved in many aspects of Ontario Hydro, there was no focus to the involvement. There was no single forum or combination of forums that looked at the overall direction of Hydro's expansion program and its need for revenue from a public policy perspective.[24]

This conclusion is even more obviously true of the energy field as a whole.

In this connection we note that a report prepared for the Organisation for Economic Co-operation and Development has come to very similar conclusions to our own. It identifies the main components of energy policy-making in Canada much as we have done and its assessment of their historical development underscores our own. It emphasizes the quality of secrecy and pseudo-objectivity in which decisions tend to be made and defended, citing particularly procedures in the federal cabinet, the National Energy Board, the Atomic Energy Control Board and Atomic Energy of Canada Limited, Ontario Hydro, and the Ontario Ministry of Energy. The report emphasizes the 'unclear' relations between the latter two and the Ministry of the Environment as major obstacles to coherent energy policy formulation and for-mal piecemeal and restrictive processes as a major obstacle to meaningful public participation.[25]

The major challenge that faces us in the energy field, then, is not 'to meet the demand' or even to minimize the 'side-effects' of energy exploitation and use. The real challenge is a political one: to recognize the social implications of energy policy and to create institutions capable of dealing not only with the technical problems, but also with the social ones in a truly democratic and open fashion. Our brief review of the present institutional arrangements does not hold out a strong hope that these goals will be achieved in these circumstances.

INSTITUTIONS AND SOCIAL POLICY

In the existing institutional setting it will come as no surprise that much of the present initiative to develop and debate full-fledged energy policies is coming from a loose collection of relatively informal private institutions, including the Canadian Arctic Resources Commission, the Solar Energy Society of Canada Incorporated, Energy Probe, and Friends of the Earth. These are only a few examples among many hundreds of national and regional organizations and initiatives that exist in response not only to special policy interests but to yawning gaps in the institutional means for initiating public discussion of policy in Canada.[26] Future energy policy institutions must be capable of perceiving these institutional gaps, of moving to fill them, of encouraging others to fill them, and of co-ordinating the work of all.

We do not pretend that it is an easy matter to say what the institutions capable of resolving these problems and of playing the four roles outlined should be like. It seems necessary to study each situation and develop proposals specifically suited to it. While it is beyond the scope of this book to produce the kind of detailed institutional proposals that are required, in what follows we provide some ideas for the direction in which more detailed studies might move.

Informal institutional realities
Consider an Ontario Hydro double its size managing a largely nuclear system three or four times bigger than now that is of absolutely crucial importance to every activity and that requires tight control, unquestioned obedience to decisions, and exclusion of the public for safety's sake. The decision-makers for this kind of system are necessarily few as the coherence of a centralized system demands this. They will have near absolute authority, for the safe management of an inherently risky, extremely sophisticated technology in a centralized control setting demands this. They will, inevitably, have similar educations and approaches to decision-making, and they will spend most of their working year dealing with a handful of similarly minded people in the various nuclear industries and their government regulatory agencies. Indeed, their jobs largely turn out to be interchangeable and personnel are regularly interchanged. These are typical conditions under which there emerges a strong group identity with a common outlook, convinced of the correctness of their preferred procedures and tending to dismiss alternatives. When so much responsibility rests on their shoulders there is naturally enough a strong resistance to any 'interference' in internal planning that is made more effective by the technically complicated nature of the technology. The typical result is a 'runaway' institution, pursuing its own version of human welfare, relatively impervious to outside influence. So far as the life experience of the large majority of people is concerned, a society filled with such institutions working in even modest collusion would severely limit the control people have over their own lives.

There are, it seems, two broad modern paths to an authoritarian hell: the one by the emergence of some central state dictatorship, right or left, by choice or revolution; and the other by allowing centralized technological bureaucracies to escalate unchecked in a political or institutional vacuum. While Canadians have shown themselves refreshingly aware of these deep political issues in such areas as civil liberties, they have revealed a dismaying complacency over the lack of freedom of access to information and the steady concentration of political power in the federal cabinet. Canadians have scarcely begun to appreciate the similar issues to energy policy-making.[27]

In short, it is not enough to place formal requirements on institutions, assigning them certain goals and responsibilities through legislation; it is equally necessary to take carefully into account the informal dimensions of the institution. Much of what an institution does and how it does it, and much of what it is capable of doing, cannot be found out by examining either its enabling legislation or its individual job specifications. These matters can only be discovered and understood by taking into account the actual social roles of the people filling the formal roles. Too much discussion of energy policy-making processes (and of other policy-making processes as well) largely or entirely ignores the informal dimensions, to its cost.

This leads to another general point of importance in the energy field: There are long-term social consequences to energy policy that come not from the energy technologies themselves, but from creating the institutions necessary to formulate and execute the policy. These social consequences may be of the first importance to take into account when assessing the policy, yet they can usually only be understood by carefully examining the likely informal roles and effects of the energy institutions in question.

Adequate institutions
We assume that what most people want is an 'open' society where there is vigorous public participation in decision-making, where personal freedom is maximized in a way compatible with collective restraints, and where social institutions co-operate to achieve a high quality of life for all. Certainly we aim for this ideal community. It is well known that theoretical ideals are not attainable; the question is what works, what will most effectively move us in that direction? So far, we have entered some cautionary notes concerning the kind of institutional solutions to look for. Now we attempt some positive suggestions.

First, there are some general desiderata on adequate energy policy-making institutions.
- They should be capable of formulating public policy in a total societal framework.
- They should be capable of fostering vigorous public debate concerning alternative public policies.
- They should have available to them some means by which the debate can be resolved in favour of a specific policy in a democratically sensitive fashion.

- They should be linked to other policy-making institutions and to parliament so that the integration and democracy called for are institutionalized in practice.
- They should have the legislative power, financial backing, and administrative capacity to enact the policy.
- They should be able to constantly reassess the chosen policy as circumstances unfold and be able to re-open the debates, or have them re-opened by other interested parties, at any appropriate time.
- They should be flexible enough to redesign their own structure and function.

But saying all this leaves the situation ambiguous because the informal roles of the institution have been omitted. For instance, one of the ways to understand these seven criteria for the energy field is by describing a powerful, centralized Ministry of Energy, rather like our present ministries, or Ontario Hydro, but with wider expertise and some more powerful management procedures built in. But this interpretation of the guidelines does violence to the intended meanings of such terms as 'democratic' and 'flexible,' though it is well known that people are quite capable of reading formal requirements in predetermined ways. Enough has been said to indicate why we believe that the informal dimensions of this institution would in the long run defeat its initial ideal.

We move towards a more definite conception of desirable institutional structure by adding that policy-making processes should be learning processes for the whole community. Policy choices present opportunities for a community to learn about its biophysical, social, and moral structure and function and present opportunities for each individual community member to learn about the needs and aspirations of his fellow members and of constructive ways to resolve conflicts among them. All social learning involves experimentation with different institutional arrangements and with different technological systems. There are, however, also strong reasons that mitigate against precipitating ourselves into a discontinuous and haphazard succession of social and technological experiments. Not only would such a policy be extremely wasteful of resources, it is also likely to be destructive to society as well as to the natural environment. Further, if both social continuity and social experimentation are desired, then regional diversity must be introduced. For in this way, and essentially only in this way, can many different approaches be tried while subjecting those living in each region to minimal stress.

With respect to energy systems, for example, we can contemplate the situation where different regions of the province choose to experiment with different energy technologies as dictated by the local geography and social aspirations. In some regions hydro-electricity might dominate; in other areas a mix of small generators (gas, coal, peat, wind powered), bio-digesters, and co-generation might prove most acceptable. Such regional diversity provides a natural basis for, and works most effectively with, a regionally diversified system of decision-making.[28]

We also have ample evidence to suggest that meaningful participatory control has to start at the local level. People who are not granted much say in the organization and management of their own communities usually do not take an interest in more general political concerns. Democratic political action consists largely in conflict resolution, and conflicts are most easily and most creatively resolved in a local setting where there is ample occasion for discussion and argumentation and also for the development of shared perceptions and goals and for mutual respect and trust. If this aspect of social interaction is missing at the local level, we can hardly expect it to emerge at the more abstract and complex level of interregional politics.

But there is more to the social advantages of local institutions than more effective participation; there is the development of an informal cultural structure that is essential to a rich human community. We can hope that safe-fail designs of local institutions will encourage them to develop an ethos in which, not success alone, but the intelligence of the experiment and of the response to failure count as criteria of competence. This in turn opens the way to a widening scope for a kind of leadership that genuinely serves the community.[29] This is an old idea, much abused; its value may be more reliably realized in the kind of local setting we are discussing here.

Decentralized institutions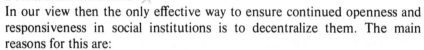
In our view then the only effective way to ensure continued openness and responsiveness in social institutions is to decentralize them. The main reasons for this are:
– There is no other effective way known to us to ensure, working within limited human capacities and resources, that there is adequate awareness in the centralized institution of the variety and detail of lifestyle and aspirations in the society and of the creative resources informally available to further its policies. For example, consider differing aspirations between rural and inner city dwellers and between northern and southern communities. Consider also the informal resources base for the development of solar and wind technologies.

- There is no more effective way known to us to promote a wide variety of detailed formal and informal interrelationships among institutions in a manner that promotes learning than by having many institutional representatives meet together in the context of the real 'field' situations with which their institutions are dealing.
- Face-to-face dialogue is much richer in possibilities for feedback, criticism, goal modification than formal communications emerging from disparate situations. In any case, the variety of actual informal relationships cannot in fact be managed to any large extent and the attempt to do so would itself create important institutional violations of freedom so that maximum richness among these contacts must be promoted.
- Local, small-scale institutions can show more flexibility and immediacy in the resolution of conflict than massive centralized bureaucracies. There is more opportunity for informal compromises, goal adjustments, and individually tailored responses than in large, formal institutions where complex formal and usually adversarial decision procedures predominate. Properly designed, decentralization may enhance democratic efficiency, since only those affected by a decision need participate.
- Decentralization allows simultaneous small-scale experimentation and learning without immediately facing large-scale consequences, thereby providing more continuity and long-term choice than centralization.
- Decentralized institutions allow for maximum learning on the part of citizens.

For all these reasons then we propose a principle of decentralization: to the maximum feasible extent, make every decision local. We believe that decentralized policy-making institutions are essential to break out of the culture/lifestyle/resource/environmental traps modern societies find themselves in. There is no 'technical' solution to most of these problems, for large-scale technological responses following our existing technological policies lead to corresponding larger social and environmental costs. In our view, our generation and the next will have to be able to evaluate choices of human value and lifestyle along with narrower economic ends if western societies are not to enter a degenerating technical-response/social-cost spiral.

Note that, unlike some, we do not argue against technology in general as if it were intrinsically evil. Rather, we argue for appropriate technologies, for technological designs that are compatible with our most valued ways of life. Indeed, we hope for sophisticated technologies to come to our aid as we face energy shortages; for example, for solar and biomass technologies to support institutional decentralization and for new communications technologies to reduce the need for centralized transportation systems.[30]

We recognize, however, that there are real limitations to decentralization. The most important of these arises from the very interdependencies we have emphasized. Material and social resources are not distributed evenly, on either a local or planetary scale; cultural diversity is also largely spatially distributed. Thus there are two fundamental arguments for society-wide institutions: every public action affects people differently and a responsible society must choose policies in a way that maintains a high quality life for all while preserving diversity; societies can no longer claim to be globally independent, since the fate of each is increasingly closely linked to that of others. Common humanity and moral integrity demand that a society play a responsible role in the planetary community. In addition, there is this consideration: given the weakness of global institutions and the adversarial character of many global relationships, a society must protect itself from global rapaciousness. These three fundamental conditions of present life all press the importance of coherent or integrated policies. It does not follow, of course, that centralized authoritarian institutions are required; what is required is that there be institutions capable of resolving regional conflicts and of formulating coherent foreign policy on behalf of the regions.

In addition to these important reasons for broad institutions, two more arise out of the potential dangers of decentralized institutions. First, there are many problems that may emerge in face-to-face settings that require larger institutional resources for their resolution.[31] Second, decentralized institutions face dangers such as local prejudice and corruption and destructive regional competition. Avoiding these means balancing decentralization with larger institutional resources. However, we suggest that the dangers of decentralization are much more readily seen and dealt with than are the corresponding dangers of centralized control. It is easier to create social authority or privilege than it is to eliminate it.

Because we adhere to this order of priorities we are cheered to discover that there is at least one energy policy direction that is thermodynamically, environmentally, and socially adequate and supports a decentralist policy: namely, the conserver society discussed in Part III.

LOCAL ENERGY OFFICES

What sort of specific decentralized institutional arrangements suggest themselves? There is no one answer, and anyway answers cannot be known in advance of experimentation. But there has in fact been some pertinent experimentation. The Man and Biosphere programme leading up to the 1972 United Nations Conference on the Environment in Stockholm was a local, participatory programme that, by and large, successfully involved thousands in a learning, values-clarifying experience. Unfortunately, from our perspective, the programme did not lead to any permanent, locally grounded, participatory roles in environmental decision-making. More important was the federal government's Eneraction programme under the auspices of Energy, Mines and Resources Conservation and Renewable Energy Branch, wherein local groups in four provinces were formed to develop and implement energy savings. Many of the local groups were extremely happy with their activities, some lasting achievements resulted, and several thousand people became informed on energy matters. Unhappily, there was evidently a certain amount of friction between the federal bureaucrats trying to control the programme and those at the local level and the programme was halted after eight months. Although there may have been specific occasions for concern given by both sides, the lesson we draw is that local groups must be given much more supportive and flexible backing by centralized bureaucrats, and central bureaucrats must be more willing to share responsibility if the local initiative is to be permanently tapped.[32] There is at this stage in Canada's political development a dearth of initiatives to involve municipal and regional officers meaningfully in development of, and experimentation with, proposals for institutional structures for the local management of energy policy.

Lest anyone think that our argument for decentralized institutional development fails for want of our presenting concrete ideas, we offer the following brief remarks to stimulate thought and discussion. Consider the creation of local energy offices to be responsible for surveying the potential local energy resources and local potential for energy conservation; liaison with other branches of government at the local level with respect to the mutual interaction of energy and other social policies; formulating and publicly presenting a variety of local energy policies; initiating public debate on, and co-ordinating public choice of, energy policy; carrying out publicly chosen energy policy, including the stimulation and co-ordination of local energy technology research and development. Finally, they are to provide the basic components of the provincial energy policy institutions. In this latter role they would be responsible for interregional liaison, provide relevant information to provincial institutions, provide the basic membership of provincial energy decision-making councils, and co-ordinate communications between region and province to facilitate integrated decision-making. These are all important social responsibilities; taken together they would make the local energy offices among the most important of our social institutions. Let us therefore review their responsibilities in a little more detail.

Surveying local energy potential
Each local unit will have to know its local energy generation potential, particularly from solar, hydro-electric, biomass, and wind sources. Much of this can be assessed through the co-ordinated efforts of local institutions – for example – engineering, environmental, and architectural firms – and local public utilities. Similarly, the opportunities for local energy conservation through the use of co-generation schemes, seasonal heat storage systems, urban redesign and retrofitting schemes have to be known if an intelligent choice of local energy policy is to be made and intelligent support given to local industry.

Local government liaison
There is a need for careful co-ordination among various social policies if a coherent energy policy is to emerge. We assume that in the long run the other major areas of social policy will be represented at the local level in a similar fashion to that for energy and suggest a permanent secretariat within the local energy office specifically to be responsible for co-ordination and balanced public presentation. Besides the divisions of local government itself, the secretariat could be expected to maintain similar contact with local industry. We suggest an industrial energy council for this purpose composed of representatives from local industry and commerce, the largest energy-consuming public institutions in the region, and members of the secretariat.

Finally, it would fall to the secretariat as well to maintain contact with neighbouring energy offices. Many issues that cannot be settled purely locally could nonetheless be settled by a few neighbouring regions without it being necessary to make the matter a provincial issue. We anticipate an interlocking network of interregional committees, formed *and dissolved* as the occasion demands, to handle such issues as the management of the entire watershed of a local river passing through several regions.

Formulation and presentation of policy
Salient local policy alternatives must be presented in a carefully integrated fashion with their full social implications made clear. This is a matter of intuition and judgment, not computation, and its morally responsible execution requires integrity, responsibility, and wisdom. Since these qualities cannot be programmed into an institution, society requires critical control of the process of formulation of the alternatives; in particular, mutual criticism among the various groups associated with the energy office will undoubtedly prove important.

To present policy alternatives to the public effectively, the energy offices will have to have adequate access to the communications media. They will have to be able, for example, to initiate newspaper, radio, and television discussions of policy. Moreover, we envisage much more vigorous public communication in the future than is now available – such as 'two-way' home-based information display and voting schemes – so there will have to be a stronger connection between the energy offices and communications authorities than that of mere buyer of programme time to media owner. We might expect a joint communications committee operating between each energy office and the local educational television authority and the Canadian Broadcasting Corporation.[33]

Public participation
The local units will initiate public debate and co-ordinate public policy as well as elicit public opinion through such means as formal referenda and telephone surveys. We suppose that focus on local versions of all issues, employing locally available institutions, may allow the time, flexibility, and dialogue that consensus-developing decision processes require.

Effecting energy policy
The traditional core of executive responsibility includes such activities as drafting regulations and penalties, licensing, policing, and subsidizing. These activities are included in the office's responsibilities, and we postulate an enactment group within the office to carry them out. But we have deliberately construed executive responsibility widely to include also responsibility for the encouragement of all the necessary supporting programmes associated with a flexible and socially responsible energy policy. Chief among them is support for selected energy technology research and development. Many, though perhaps not all, of these activities can be delegated to local enterprise and educational institutions, with the energy office retaining a co-ordinating/subsidizing role. There will need to be close co-operation between the energy offices and both provincial and federal government agencies. Moreover, we assume a close connection between this group and the local industries energy council.

REGIONAL RESPONSIBILITIES AND LOCAL INSTITUTIONS

This brief review of one suggestion for local energy office responsibilities has tacitly revealed a number of other implications for the local institutional framework and we now comment briefly on some of these.

There will need to be close liaison between the energy offices and local governments. Moreover, in a situation where each of the other provincial government responsibilities is locally represented in a similar form, local government itself must adopt more of a co-ordinating role if it is to remain relevant and if its present functions are to be smoothly integrated with these policies. Many provincial ministries now maintain regional offices that could be useful building blocks for a more effective local framework. A natural response to the creation of local ministry offices would be to install local government as the co-ordinating council of these offices, integrating their present responsibilities into this role. And it would be appropriate to enlarge the local financial base and increase its legal powers and responsibilities, including making all co-ordinated agencies responsible to the local government, at least in local matters.

If local government is to deal effectively with local policy, the spatial extent of its responsibilities must be suitably defined. Municipal governments now have responsibility for geographic areas that are, on the whole, too small and more or less arbitrarily chosen as far as the important physical/biological parameters are concerned. For example, environmental control of rivers and water usage involves policies for entire watersheds, and effective control of urban development encompasses policies for the immediate hinterland as well. Such considerations suggest that local governments be coalesced into regional governments whose jurisdiction is chosen to correspond to salient policy boundaries.[34]

To provide a rich setting for policy discussion it is necessary to have a neighbourhood base. This is in any case the appropriate scale for many other public functions and we envisage communication centres based in neighbourhood resource centres that might include a library, government services offices, integrated family health care clinic, and recreation and cultural centres. The aim would be to make them a focus of neighbourhood activity in a setting that encouraged participation and where integrated, two-way communication was possible. Within this setting the communications centre would include the presentation and debate of energy and other social policies.[35]

Finally, we propose absorbing the local electric utilities into the energy offices as a part of an energy production and distribution group. The argument in favour of such a proposal is that the production and distribution of electricity can be made local policy issues in the way that they cannot wholly be for 'conventional' primary energy fuels and that with increasing emphasis on seasonal storage and mixed supply systems they will become an increasingly tightly integrated aspect of local energy policy. Moreover, some energy offices might, for diverse local reasons, choose to experiment with their own

FIGURE 17

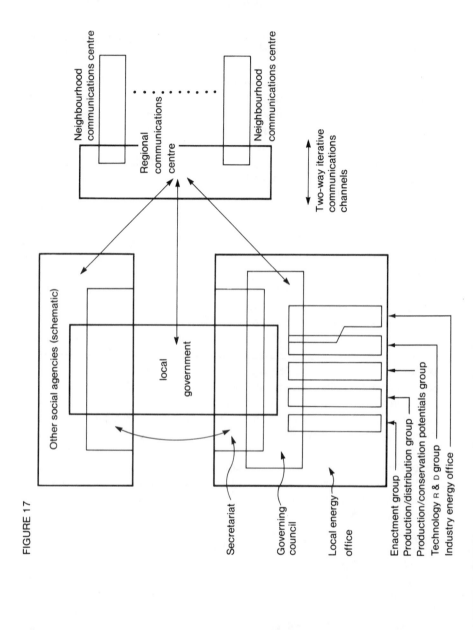

Neighbourhood communications centre

Neighbourhood communications centre

Regional communications centre

Two-way iterative communications channels

Other social agencies (schematic)

local government

Secretariat

Governing council

Local energy office

Enactment group
Production/distribution group
Production/conservation potentials group
Technology R & D group
Industry energy office

FIGURE 18

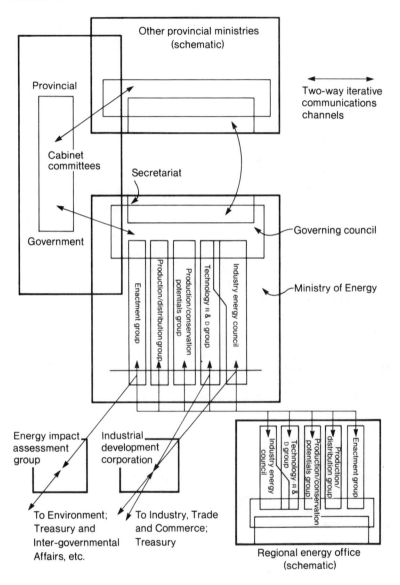

Two-way iterative
communications
channels

Governing council

Ministry of Energy

Regional energy office
(schematic)

investment in energy technology, and particularly 'alternative' technologies. In this way they might acquire other direct responsibilities for the production and/or distribution of energy. All of these responsibilities are assumed collected in an energy production and distribution group.[36]

The picture of local energy offices structure we have built up is schematically represented in Figure 17. The overlapping areas indicate some common membership. We might further envisage a provincial ministry of energy of similar structure and function as in Figure 18. Attached to the ministry of energy we suggest two further, new institutions. First, a provincial energy impact assessment group, conceived to operate in a fashion parallel to environmental impact assessment groups. Membership in this group would be from all relevant ministries as the occasion warranted, and its function would be to assess the *full* range of potential effects of specific energy policy proposals.[37] Second, we propose the creation of a provincial energy development corporation to promote the development of new energy technologies by a combination of subsidizing special development programmes in existing industries and initiating entirely new industrial enterprises where necessary. The point is to ensure that new, desirable energy technologies are not rejected simply because the past pattern of industrial investments does not make them economically attractive investments at the present time. We assume strong regional representation in the corporation, and one of the corporation's most important functions will be to support local choices of energy technologies. The present Ontario Energy Corporation, which has concentrated on investment in existing fossil-fuel supplies and technology, may need to be retained, perhaps with a modified direction such as towards development of small, clean, efficient coal-fired electric generators. Whether the two corporations should be merged or left separate, we do not attempt to decide here.

In the figures we have specified neither the precise details of common memberships nor the structure of the iterative processes of decision-making to be followed. Such a detailed proposal, if it were to have any hope of leading to practically workable institutions over the long term, would need a background study many times larger than the constraints on this study permit. Indeed, further reflection may well uncover many difficulties even with the general model outlined here. We have only suggested a model to illustrate our ideas. Our main concern is that there should be an immediate beginning on a serious study and public debate of these alternatives.

In Figures 17 and 18 we indicate only a few examples of the many communications channels we think are necessary to maintain an open, consensual institution.[38] We suggest that all discussions be made public unless there is an explicit cabinet decision to the contrary based on justifiable and fully explained grounds and that the regional communication groups be explicitly charged with opening these discussions to local public debate. But communications alone do not settle all aspects of local-provincial relations. A viable decentralized society must have at hand a set of local-provincial co-ordination and conflict resolution procedures that strike a balance between the legitimate role of local and society-wide factors while safeguarding society from the defects of either level of decision-making pursued in isolation. It is really beyond the scope of this study to offer detailed institutional models in this difficult area; in any case, they should properly emerge from wider debate and experimentation than any small group can provide. In what follows we discuss one proposal for developing a new class of conflict – resolution institutions.

CRITERIAL INSTITUTIONS

An example of the institutional process we have in mind is provided by the Ontario Municipal Board (OMB), the last court of appeal for local decisions concerning land use, construction, and related matters. Institutionally, it is neither a part of the judiciary nor a part of the civil service immediately attached to some ministerial portfolio; yet it is a public institution with quasi-judicial powers. We refer to it as a 'criterial' institution because its decision-making processes focus upon the attempt to apply not simply a set of regulations, but a broad set of criteria. By most accounts the OMB has worked well and certainly been an improvement over what the situation would have been in its absence. But it is not so much the specific details of the OMB to which we wish to appeal. Rather we draw attention to the general idea that there could be public institutions with conflict resolving roles that lie in a more 'neutral' space between the judiciary and the normal civil service.

In brief our suggestion is that the imaginative creation of criterial institutions may be an important factor in achieving adequate local-provincial relations. At present we are faced with a hierarchically ordered, moderately secretive, centralized civil service and weak local governments. A primary aim of decentralized institutional design would be to strengthen the local institutions and free up the provincial decision-making process without sacrificing provincial coherence and good sense. Criterial institutions have the intrinsic ability to operate within a provincial framework, yet have a certain independence of special-interest civil service procedures. And they could evidently be designed to foster an open exchange of views in a framework that focuses on the evaluative bases of social decisions. They might therefore prove an important means to structuring local-provincial conflict resolution processes in a desirable, yet controllable manner.

The OMB is not the only example of a criterial institution. The federal Central Mortgage and Housing Corporation shares at least some of the same features, as does the Canadian Radio and Television Commission. In the energy field, even Atomic Energy of Canada Board, and, for that matter, Ontario Hydro, share the same non-judicial, non-civil service status and possess regulatory powers. So the idea of creating institutions of this general type for quite specific social purposes is not new. We only suggest exploring a more imaginative, 'fine-grained' set of structures of the same type. In this respect the new Institute for Man and Resources in Prince Edward Island constitutes a more interesting institutional model in the energy field than does Ontario Hydro because of its greater integration and greater community involvement. Note, too, that none of the other institutions mentioned here shares with the OMB the responsibility of holding open hearings and of reaching decisions based on the arguments presented. There is, however, at least one energy body in the province that has an open hearing role: the Ontario Energy Board. Though this body lacks the requisite relations with local government that we envisage, and largely lacks relevant conflict-resolving powers, the hearing process does provide some opportunity for genuine interactive, iterative learning in the energy field. Royal commissions, for example the Royal Commission on Electric Power Planning, have a similar capacity to initiate public discussion. Thus there is a precedent for a participatory approach.

Once again we stress that we do not propose criterial institutions as a panacea for the complexities of the energy policy field. There must be many other dimensions to the local-provincial institutional processes if the result is to be viable, and other institutional arrangements may prove more valuable models for these. There may, for example, be value in comparing the Ontario Hydro – local utility relationship with the Ministry of Education – local school board relationship when looking for fruitful provincial-local co-ordination and planning relationships, as well as when looking for difficulties to avoid. Again energy is coming increasingly to look like a fundamental social service, so far as public policy is concerned, because of the widespread social ramifications of energy policy decisions. In that case there may be equally useful insights to be had from the comparative study of local-provincial institutions for the management of public water supplies, health care delivery, environmental conservation, and even fire and police protection. We suggest that there are valuable insights to be had from the Yugoslavian experience in the local-national co-ordination of workers' councils that have some executive control of production decisions and from the experiences of the Danish energy ministry in the development of local, renewable energy policy.[39]

Having sketched a possible provincial structure for energy policy-making institutions, there remains the question of the place of Ontario Hydro in the scheme. We contemplate that the policy-making sections of Hydro would be absorbed into the provincial-regional structure, since that would properly be their responsibility. In particular, this would include the load forecasting personnel and system expansion and investment personnel. These skilled personnel would easily find appropriate places in several groups and councils of the ministry and regional office structure and would be invaluable in the initial stages. As for the remainder of Ontario Hydro, there are considerations that favour retaining it as a commercial operation and some that favour dispersing it among the local-regional structures. We do not attempt to weight these here.[40]

Decentralization would require substantial institutional change since present energy institutions in Ontario are a long way from it. There will be a natural resistance to these changes. Though tempting, it will prove important not to build solely on the basis of existing institutions but to ensure that they are subordinate to the alternative design. Proceeding slowly, say over fifteen years, and employing a judicious combination of modified existing institutions and new ones, it should be possible to develop a new structure without serious disruption either to people's personal lives or provincial energy policy.

At present it is important to experiment with these ideas. We suggest that a few regions of Ontario be designated as experimental regions and that the government create local energy offices, with a public timetable for giving them powers appropriate to their functions, say over three years, and then a specified review process after a fixed time, perhaps in a further three years. The review should be wide ranging, carried out by a balanced group including representatives of business, labour, citizens groups, and provincial and local governments, and be publicly conducted and reported. The acknowledged aim would be to build wider experiments on the experience gained and the final, public report should make recommendations for the next round of experimentation. We suggest that such a programme of institutional innovation receive the same priority as favoured technological innovation ought to receive.

PUBLIC INSTITUTIONS AND PRIVATE FIRMS

We turn finally to the last group of energy institutions not addressed in these proposals: the private energy firms. Our basic philosophy is that the market is an effective way to allocate resources within certain social limits. Once the social limits to each commodity market have been carefully set and questions of income and wealth distribution addressed, both social freedom and economic efficiency are promoted by letting the market operate within these constraints. This is the basis for our twofold suggestion concerning possible working relations between private and public energy institutions.

First, the relations between these firms and federal and provincial governments would continue to develop along present lines, with some important modifications. Historically, federal and provincial governments have come to play increasingly active roles in the regulation of the activities of private energy firms. They exercise this role through their influence on pricing, economic investment incentives, export and exploration regulations, pipeline and refinery construction decisions, environmental regulations, and so on. We suggest a continuation of this intrusion, though in a context offering firmer, long-term expectations more influenced by government-generated information and publicly discussed local energy decisions.

Second, at the local level the behaviour of local representatives of these energy companies would conform with constraints set by local energy offices. Since most of these constraints will follow naturally upon decisions concerning energy production technologies, we envisage the focus of decision-making to be on them and we expect the need for local regulation of fuel sales and equipment servicing to be minimal.

To implement these relationships would involve a substantial transformation of the historical situation of an energy market dominated by private companies acting, in the main, only within its constraints. Even so, we do not believe that there must inevitably be a winner/loser conflict situation. Our society has come increasingly and properly to intrude its larger social interests into the energy market as the social and economic consequences of energy decisions have become more apparent. Private energy companies have learned to live with that intrusion and there is no reason to believe that this will cease to be the case in the future. In addition, we must unavoidably become substantially independent of fossil fuels and (unprocessed) uranium in the relatively near future and this inevitably means major changes for the majority of private energy companies. Thus private energy companies are already bound up in a process that will transform them or see them lost by the way. The transformation of the role of private energy companies is a desirable and largely unavoidable process in which it is to be hoped that public institutions will play an active and constructive role. Finally, there are now many industrial opportunities in the new energy technologies and energy fuels and there will be many more such opportunities as the future unfolds. Putting these three factors together we conclude that there is every reason to believe that even within a strengthened public decision-making framework there remains a viable, indeed a substantial, role for private enterprise. This 'substantial role' will be open to those private enterprises willing to adapt their own institutional structures to the emerging social context in which they will be operating. Rather than a period of conflict, the transition in the next forty or so years may be approached as an opportunity to work out new private-public relationships characterized by interaction and co-operation.

LOCAL PARTICIPATION AND PARLIAMENTARY DEMOCRACY

We have now discussed all of the categories of specifically energy decision-making institutions, but we have not yet considered the future role of the most general decision-making institution in a democracy, parliament. It would be contrary to the whole drive of our discussion of institutions, which has been in support of democracy, to neglect the role of parliament; yet the preceding discussion may have left the impression that there is little role left for it in policy formation. And the fact is that parliaments are already under siege. As the number and complexity of decisions increase apace, more and more decisions are effectively made outside parliament and beyond the knowledge and competencies of parliamentarians.[41]

Whatever the future of parliament, if genuinely decentralized decision processes can be established, the result will be positive because decision-making power will spread and the flow of information open, reinvolving the public in taking responsibility for their lives. In this way genuinely democratic processes will be supported and hence serve the ideals for which parliament was intended. But it cannot be denied that institutional developments of the kind we discuss would spread decision-making power well beyond the confines of parliament and hence challenge its exclusive role in policy formulation. This is no more than an extension of what, for example, already happens through the Ontario Municipal Board and the civil service. In these circumstances there are at least two pertinent responses.

First, the role of parliament can shift to increasingly higher levels of decision-making, to focus on matters of principle, large-scale socioeconomic policies, and innovative legal development. In fact, the trend in this direction is already strong. Second, parliament can become more actively involved in the decision-making process through increased roles for parliamentarians in the decentralized institutions. We might, for example, consider establishing a minimum parliamentary term but require that part of this be spent actually serving various of the policy-making institutions in some capacity. Or we might consider extending the system of parliamentary committees so that they become integral to the extended decision-making process. In fact a combination of both transformations of parliament applied together strikes us as having the attraction of balancing one another, without cancelling one another. No doubt all such proposals have disadvantages and require much careful thought before enactment. The point is that the present form and role of parliament can be extended in a variety of ways, and if this is done wisely a better informed, less posture-ridden parliamentary process is likely to result.

A BUREAUCRATIC FIX?

Some readers may react to these institutional proposals by arguing that they represent a 'bureaucratic fix' stemming from the naive belief that switching to yet more bureaucracy, but of some 'ideal' design, will realize utopia. As such, the argument might run, our proposals are as bad as the 'market fix' that sees the market as solvent for all social conflict, and the 'technological fix' that relies on more sophisticated technology and technologists to save us all. Worse, our suggestions may be thought to represent an alternative that will inevitably degenerate into rule by secretive bureaucratic authoritarianism, an end we have deplored as much as its opposite extreme, market incoherence. Public participation, never strong in such matters, will not be strong enough to prevent such a situation.

There is a real difficulty represented by this response. But before coming to that we cannot resist replying that the institutions we propose are hardly likely to be less participatory or more authoritarian than our present energy institutions. Indeed, since over half of the personnel now employed in the nuclear field are civil servants, following a decentralized, renewable energy future might make a substantial contribution to reducing the existing centralized bureaucracies.[42] Nor have we advocated abolition of markets, only social control of their scope. We are pluralists, and private enterprise provides an important counterbalance to state power. Local participation and autonomy provide another.[43] This said, let us turn to the central thrust of the objection.

The objection really gathers its force from the tacit assumption that there are basically only two alternatives that define the spectrum of available societies: free-market democracy at one extreme, and state communism at the other. There is indeed some plausibility to this assumption in our present circumstances. The major international competitor to capitalism is communism and in all its major instances it has taken a centralist-bureaucratic form, though with wide internal variations. Under capitalism, where the market has not reigned supreme, it has been modified by the institutions of the welfare state and, as we have stressed, both the private enterprises in the market and the welfare state institutions modelled on them have tended increasingly to centralized, commodity-oriented, bureaucratic control. If we cannot move outside of this experience, then the choice boils down to just how much central bureaucracy we want and an intelligent answer surely commences with the constraint: 'as little as is reasonable.' But this lesser-of-evils way of putting the choice obscures what is at issue in being reasonable here.

What is surely at issue is the attempt to provide richly participatory, yet morally autonomous lives for all members of a society. Such lives are not reliably provided by any mixture of market competitiveness in a moral vacuum and centralized bureaucratic repression or coercion. We look instead in a third direction, towards decentralized, participatory institutions that set goals for, and constraints on, the market. In this manner we hope to retain the moral autonomy of liberalism while developing the intentional, coherent social choice aimed at by central bureaucracy – and we hope that intelligent institutional designs will succeed in avoiding the major defects of both. We cannot emphasize too strongly that this 'third way' calls for new institutional developments and that the secret to these has to lie in their ways of combining genuine local, decentralized participation with the achievement of social co-ordination. Just this, indeed, is the subject of much recent writing in this area.[44] Unless these institutions can be developed, our world of rising technical power and diminishing manoeuvrability will be in for a fearful future. The real difficulty is to develop workable institutional designs in a setting that discourages even conceiving of this option. Energy choices are an essential and obvious place to start. What is required now is commitment to the serious study of the alternatives, to public debate of them, and to experimentation.

10
Understanding energy policy

The immediate focus of our argument has been on energy policy because its implications are so fundamental, but we have tried also to bring to light the more universal human issues within this specific social issue. Briefly, in Part I we developed an appreciation of what would constitute an adequate energy policy. Since energy flows are the key to all life and industrial activity, energy policy is a basic social policy. For the same reason every aspect of the structural design of a society, from its heating fuel to its urban form and from its transportation modes to its recreation, is bound up with its energy policy. Thus the basic physical ingredient of much that determines the quality of life a society enjoys – for example, physical mobility, access to comfortable housing, and recreational options – is energy policy. It extends to the level and distribution of jobs and occupational health hazards, the structure of environmental hazards and damage, and international security. An adequate energy policy, then, is one that consciously takes all these kinds of factors into account in a coherent plan for the overall structural evolution of the society in a direction that expresses critically chosen basic social values.

Canadian public energy policy documents do not offer a comprehensive approach to the sort of energy policy we believe necessary. What is offered instead is a 'dilemma' or 'crisis' stated in terms of falling supply against rising demand and a 'business-as-usual' solution that seeks mainly to augment conventional supplies to meet the demand. Meanwhile, the demand and the social consequences of trying to meet it go largely unexamined, as do the deeper evaluative issues and hidden tradeoffs involved.

In Part II we first uncovered the elements of this business-as-usual response – basically reliance on increased oil and gas search and delivery, with expanded nuclear electric capacity – arguing that, even on its own grounds, it is so fraught with difficulty as to not present a credible approach.

Then we examined the likely consequences of attempting to meet present rates of demand growth, whatever the method, and concluded that they are overwhelmingly large and negative, including increased environmental disruption, a narrowed employment base, increased health risks, and more intensive urban centralization and concentration of power. Thus Part II closed with reflection on the ethical, social, and political aspects of energy policy. Our view is that such implications should dominate the choice of energy policy. Indeed, in circumstances in which any energy policy is faced with large technological and economic uncertainties, it is only rational to invest social resources in the most socially preferable alternative rather than, for example, the one most economically attractive in the short term.

In Part III we inquired into alternatives to the business-as-usual policy. It is clear on both narrow economic grounds as well as wider social grounds that energy conservation should be the base from which alternative policies are developed. Accordingly, we investigated the potential for energy conservation. Contrary to the slight role given conservation by governments, we concluded that the need and potential for energy conservation is in fact large. Without major lifestyle changes, for example, Canadian per capita energy demand could be halved by 2000. With larger structural changes – though not ones that would lower overall living standards and might indeed raise them – it would be possible for Canada to hold its energy demand to 1970 levels in 2025 and still allow for modest economic growth. If these conservation goals were met, almost all of the difficulties, socio-economic and environmental, associated with the present desperate search for new energy sources would vanish.

With energy demand cut through conservation it becomes feasible to consider solar-based renewable energy sources as an alternative to fossil fuels and nuclear generated electricity. Our evaluation of the potential of these resources indicates that radiant solar energy could reasonably supply low-to-intermediate temperature heat; hydro-electricity could meet the unavoidable electrical demand; and biomass conversion could fill all or most of the transportation requirements for liquid fuel. In this manner, constrained energy demands could be largely met from indigenous, renewable energy. This evaluation was not intended to fix a detailed energy policy but to indicate that conserver/renewable policies are as viable as their fossil/nuclear alternatives. The particular energy technologies chosen will depend upon a host of detailed geographical, economic, and social considerations. Whatever the precise details, we argued that the kind of conserver/renewable energy policies we explore are much more likely to promote national and regional autonomy and equity, economic development and employment, long-term flexibility and economic viability, and safety for humans and their environment.

But it is not enough to argue in favour of a general policy orientation for the energy field. The detailed decisions must still be made, and how they are made is important. Further, our analysis explicitly considers the ethical, social, and political dimensions of energy policy; these offer the strongest reasons for preferring conserver/renewable alternatives. The present institutional forms of energy policy formation have obscured or ignored these dimensions, and this error needs to be corrected by developing policy-making processes adequate to the richness of the field. Finally, and since participation is largely absent from existing institutions, this leads us specifically to espouse more participatory energy policy-making institutions.

Thus Part IV of our argument is devoted to an exploration of a system of decentralized regional energy policy-making and managing institutions integrated with a central co-ordinating institution. Again this is presented as a first attempt at an alternative to present-day centralized bureaucracies, not as an inflexible proposal. Indeed, what is now called for is experimentation and open discussion to facilitate public learning in the choice of alternative institutional arrangements.

But why has energy become an important issue in the second half of the twentieth century? After all, energy flows were as fundamental to the lives of palaeolithic man as to ourselves. An insightful answer depends upon an appreciation of the unique confluence of several facets of human development in the twentieth century.

First, and most obviously, the scale of energy use has increased. Energy resources on a planetary scale are now being consumed in a single lifetime, and man-managed energy flows now amount to a significant fraction of total planetary energy flows. The present sense of urgency surrounding energy problems is ultimately traceable to the realization that current energy use patterns will soon be constrained by planetary limitations.

Second, the increase in energy flows associated with human activity has provided a basis for the qualitative transformation of human society from small hunter and gatherer communities to large urban-industrial complexes. The highly specialized forms of social and economic life that now predominate in modern industrial societies are made possible by the energy subsidies we have provided ourselves. For example, 5 per cent of the population can produce enough food for the 95 per cent that are urbanized only if there are massive energy subsidies to the agricultural sector. Growing awareness of this intimate relation between our specific ways of life and energy flows has added a particular sense of gravity to energy policy issues.

Third, the nature of industrial society is such that the consequences of energy policy now span the globe and extend several, sometimes hundreds, of generations into the future. The exhaustion of easily accessible world oil supplies will profoundly affect the development patterns of the third world, and development of nuclear electricity will impose constraints on many future generations. It is no longer possible to ignore one's neighbours in space or future generations in time when making energy decisions. This adds a new complexity and a particular sense of urgency to decision-making before the consequences overwhelm us.

Fourth, future energy policy options have become increasingly dependent on the present social organization of scientific knowledge and research efforts. Whether photo-voltaic cells or thermonuclear fusion will provide attractive energy technologies in the future will depend heavily on the research devoted to them and on the kind of deliberate economic support provided for their implementation. Moreover, there are increasingly complex and intimate interactions between scientific knowledge organized around energy and that organized around other economic activities. In this connection, a northern Canadian biomass industry may be an important, and perhaps a necessary, factor in the development of a northern-based chemicals industry utilizing closely related chemical processes. More ominous is the clear and demonstrated link between the use of 'peaceful' nuclear technology and the capacity to build nuclear weapons – the most pressing international challenge of today.

These features of the energy policy field provide an additional dimension of complexity and social involvement to energy policy issues. They are all unique to the twentieth century and they give energy policy a special urgency and complexity. They are not, however, unique to energy policy. All public policy areas of any general structural importance – for example, communications, transportation, agriculture, education, and health – display the same features. This is what lends a sense of importance to developing a general approach to public policy-making for the era in which we live.

In fact every major public policy area exhibits analogous issues to those in energy. Consider communications, for example. First, the scale of communications has increased and at an especially rapid rate this century, and already planetary communications structures – for example, communications satellites – are coming to intrude noticeably on national cultures. These developments lend a sense of urgency to communications policy. Second, modern communications technology, such as computer and video technologies, have begun to transform the qualitative pattern of social life, releasing societies from the necessity of physical travel to communicate and from physical intervention to control. This intimate relation between social structure and communications structure adds a particular sense of gravity to communications policy. Third, the consequences of communications policy now span the globe and extend generations into the future. Modern communications have, for example, made the European Economic Community a possibility and the emergence of global satellite monitoring and communications will be felt in international relations far into the future. This adds a new complexity to communications decision-making and lends a sense of urgency to making appropriate decisions now before the consequences overwhelm us. Fourth, future communications policy options are heavily dependent on the present social organization of scientific knowledge and research effort. This is strikingly illustrated by the revolution in micro-electronics that has taken place in the last decade under a deliberate and intensive research programme, a revolution whose profound effects are only just beginning to be felt. Moreover, there are increasingly intimate links between communications research and development and other social areas, such as educational processes, industrial automation, and even political processes.

In our view these general features call for a new approach to public policy-making. The old conception of a monolithic bureaucracy concerned only for its own policy field, typically focusing on a single economic commodity and working towards the most economically efficient technology for the field, is obsolete. In the recent past, Canada had, or had fragments of, an oil policy, a gas policy, a wheat policy, a fisheries policy, a lakes and rivers policy, and a forestry policy. The lines of division followed commodities in the market, and ecological, energy, and nutritional policies were, and largely still are, absent.[1] The real consequences of decisions are obscured or ignored and in the long run this is dangerous. Moreover, the peculiar importance of the organization of knowledge to twentieth-century decisions makes the development of public policy in centralized institutions potentially harmful to political freedom and societal flexibility.

Once the special nature of twentieth-century public policy is appreciated, however, it becomes clear that what is called for is not just a statement of existing policy options in some field or an argument for the 'correct' present policy but rather a new conception of what should be involved in a policy and a new institutional setting for developing policy options and for choosing, pursuing, and revising particular policies. This is why we closed our discussion of energy policy with a discussion of energy policy institutions. Arguing from general considerations, we recommended exploring decentralized institutions with rich interactive processes. These institutions would represent the full complexity of energy policy and develop adequate policy and policy options that do not reinforce centralized control.[2] Once the particular nature of twentieth-century public policies is appreciated, the particular urgency and importance of this political/institutional issue can be properly perceived.

And here too the twentieth century has placed us in an unprecedented position. This century has witnessed the development of a unique class of theories concerned precisely with the behaviour and management of complex systems: systems theory, control theory, cybernetics, decision theory, operations research, ecological dynamics, network and bond graph theory, organization theory, artificial intelligence, and so on. Though often focused on different systems and employing different terminology, these theories are all centrally concerned with understanding and managing complex systems.[3] One of the important complex systems that may be studied using these theories is the social organization of scientific knowledge and research itself.[4] Such theories have begun to provide deeper insight into the institutional structures and policy formation processes that might be appropriate to our historical circumstances. We believe they point towards the importance of the kind of institutional structures and adequate policies we discuss; but it is the work of another day to develop their insights explicitly for the policy field. In the meantime, it is urgent that policy issues be carried to the level of political debate about institutional processes and commitments be made to experiment with more adequate institutional designs in a setting that still permits learning and adaptation.

If we choose, we can respond to the growing number and magnitude of our difficulties in the time-honoured primitive fashion: with fear, authoritarianism, and short-term opportunism. We can build for ourselves a self-fulfilling prophecy of inequity, elitism, and mounting disruption. But there is another, better, choice available.

Notes

In the following notes, italicized numbers following titles refer to the number of the complete reference in the Bibliography; RG refers to the Reader's Guide.

CHAPTER 1

1 See V.R. Puttagunta, *Temperature Distribution*, *328*; see also A.B. Lovins, *Energy Strategy*, *228*.
2 For an account of many of the present systematic energy inefficiencies see, for example, B. Commoner, *Poverty of Power*, *61* and his earlier *Closing Circle*, *57*. See also RG, G.1 and J.2.
3 See, for example, the extended discussion in C.A. Hooker and R. van Hulst, *Institutions*, *173* and RG, L.2 and L.3.
4 See W. Berry, *Culture and Agriculture*, *20*; A. Birre, *Politique de la terre*, *22*; B. Commoner, *Poverty of Power*, *61*; P. Ehrensaft, *Canadian Agriculture*, *97*; D. Mitchell, *Politics of Food*, *257*; H. Nash, *Progress as Survival*, *267*; and H.T. Odum, *Energy Basis*, *282*.
5 See Science Council of Canada, *Uncertain Prospects*, *373*.
6 On the pulp and paper industry see E.P. Gyftopoulos et al., *Fuel Effectiveness*, *139*; on agriculture, see note 4.
7 We note, however, that whether these are actually consequences of decentralization depends very much upon the cultural framework within which regional development is approached. Decentralization and regional development may be coupled to simpler, more frugal, more self-reliant lifestyles; they may also lead to wasteful regional rivalry and to increased desire to travel. This serves to remind us that these policies cannot be pursued in isolation.
8 In June 1978 the oil and gas companies were pushing the National Energy Board to permit increased exports of oil and gas to the United States because

of the temporary emergence of 'surpluses,' despite the fact that they argued for price increases only a year earlier on the grounds that reserves were inadequate. They succeeded in having exports increased. See *Globe and Mail*, October 13, 1978, and March 7 and April 12, 1979. For National Energy Board decisions see *Canadian Natural Gas: Supply and Requirements* (Ottawa: Queen's Printer, February 1979). Private enterprise succeeded even beyond the liberal constraints set by the National Energy Board, for on December 7, 1979, the federal government agreed to export to the United States over the 1980–87 period nearly *twice* what the National Energy Board had declared surplus (respectively 3.75 and 2 trillion cubic feet); see *Globe and Mail*, December 7, 1979. For the official position with respect to oil see National Energy Board, *Canadian Oil: Supply and Requirements* (Ottawa: Queen's Printer, 1978). An interested reader can follow the approaches of private enterprise to these problems through their submissions to the National Energy Board and their submissions before the Ontario Energy Board, as well as to relevant legislative committees and royal commissions.

The Canadian Nuclear Association, which represents commercial interests in the nuclear field, has been pressing to continue the demand for the production of nuclear reactors. See their report *Nuclear Power*, *43* and their submission to the Royal Commission on Electric Power Planning, *Role of Nuclear*, *44*. See also Leornard and Porteus, *Economic Impact of Nuclear*, *223*.

9 See, for example, J.M. Blair, *Control of Oil*, *23*; J. Laxer, *Energy Poker Game*, *215* and *Canada's Energy Crisis*, *216*; J. Laxer and H. Martin, *Imperial Oil*, *217*; C. Pratt, *Tar Sands*, *327*; J. Ridgeway, *Struggle to Monopolise Energy Resources*, *333*; A. Sampson, *Great Oil Companies*, *360*; and P. Sykes, *Sellout*, *405*.

CHAPTER 2

1 Energy, Mines and Resources Canada, *Energy Strategy*, *100*. Since then there have been regular annual publications by Energy, Mines and Resources Canada that update the situation in various dimensions of energy policy. See *Energy Update*, for 1976, 1977, and 1978, report EI77-2, EI78-2, and EI79-2. See also *Energy Conservation in Canada* (Ottawa, 1977), EP77-7; *Renewable Energy Resources: A Guide to the Literature* (Ottawa, 1977), EI77-5; *1976 Assessment of Canada's Coal Resources and Reserves* (Ottawa, 1977), EP77-5; and *1978 Assessment of Canada's Uranium Supply and Demand* (Ottawa, 1979), EP79-3.

2 The much-touted 1978 West Pembina oil finds in Alberta, the largest in a decade of exploration, would add only roughly three years' present national consumption to proven reserves according to even the most optimistic esti-

mates. These current proven reserves stand at something like nine years'
supply at present consumption rates. The ultimate contribution of off-shore
Atlantic deposits remains uncertain. Recent natural gas discoveries improve
this picture somewhat, adding in the vicinity of twenty-five years' supply at
present rates of consumption. However, this is not a long time from the perspective of national social policy, and we must expect that in the present
circumstances the rate of consumption of natural gas will continue to rise
substantially each year so that these figures do not represent realistic long-
term supply estimates. Contrast these relatively dismal figures with a statement by the then federal minister of energy, J. Greene, on July 2, 1971, to
the effect that Canada had at that time over 900 years' supply of natural gas
and almost 400 years' supply of oil based on 'estimates' provided by energy
industries and substantially accepted by the National Energy Board; (see
J. Laxer, *Canada's Energy Crisis*, *216*, P. Sykes, *Sellout*, *405*). An even larger
figure was suggested by James Gray, vice-president of Canadian Hunter
Exploration Ltd, who stated more recently that Canada has a vast natural gas
potential and 'enough oil ... to supply us between 1000 and 2000 years'; see
Toronto Star, April 12, 1979. See also note 8, of chapter 1.

3 OPEC includes chiefly Middle East countries but also Venezuela from whom
   most of Canada's imported oil comes.
4 See Energy, Mines and Resources Canada, *Energy Strategy*, *100*, p. 113 and
   G.M. MacNabb, *Canadian Energy Situation*, *238*.
5 See the statement by R. Toombs, Energy, Mines and Resources Canada,
   Petroleum Supply and Demand, *419*, p. 3, to the effect that additional resources
   might be generated by increased exports of newly discovered gas to the
   United States (cf. note 8 of chapter 1). But to that extent, Canadian reserves
   are depleted and the likelihood of a supply/demand imbalance beyond 1990
   increases. Moreover, not only is the international picture rapidly worsening
   from a western economic viewpoint, not least because of rising western
   demands in the face of dwindling reserves, but there is also substantial reason
   to expect a rise in European Communist bloc demand for Middle Eastern oil
   beyond 1985 that will further exacerbate world supply shortages.
6 See, for example, B.W. Erikson and L. Waverman, *Energy Question*, *112*;
   A.R. Flower, *World Oil Production*, *114*; S.S. Penner and L. Icerman, *Energy*,
   317; L. Rocks and R.P. Runyon, *The Energy Crisis*, *340*; L.C. Ruedisili and
   M.W. Firebaugh, *Perspectives on Energy*, *357*; C. Watkins and M. Walker, *Oil*,
   445; C.L. Wilson, *Energy*, *459*.
7 See Science Council of Canada, *Uncertain Prospects*, *373*.
8 See J.E. Gander and F.W. Belaire, *Long-Term Energy Assessment Programme*
   (LEAP), *123*.

9 *Ibid.*

10 See C. Marchetti, *Energy Substitution Models*, *242*; see also P.L. Cook, *Energy Policy Formulation*, *65*.

11 See Figure 1.

12 Government policy, understood to mean the behaviour that actually affects what is happening in the country, is often illusive, largely locked away in the unrecorded informal communications of cabinet ministers, backbenchers, senior civil servants, members of the National Energy Board, and so on. None of this informal activity is directly accessible to the policy analyst in any systematic manner. What is accessible, however, are the formal documents that governments and government departments issue from time to time and to which appeals are made when matters of controversy arise. In what follows we confine our discussion of government policy to what may be inferred from the formal documents made publicly available – a justified approach in a democracy.

13 See Energy, Mines and Resources, Canada, *Energy Strategy*, *100*, p. 106. Further, 'the mid-1977 estimate for the year's electric utility capital expenditure was $5,500 million ... This represented 18.9 percent of total business expenditure for Canada,' Energy, Mines and Resources Canada, *Energy Update, 1977* (Ottawa, 1978), p. 31. These figures refer only to expenditure on *electrical* energy and do not include $90 billion or more to be expended on foreign purchases of oil. They are probably also conservative.

14 Energy, Mines and Resources Canada, *Energy Strategy*, *100*, chapter 4.

15 See, for example, Energy, Mines and Resources Canada, *Energy Strategy*, *100*, p. 83, and G.M. MacNabb, *Canadian Energy Situation*, *238* and Figure 7. See also note 2 of chapter 2.

16 See G.M. MacNabb, *Canadian Energy Situation*, *238*, p. 8.

17 See, for example, the discussion in R. Toombs, Energy, Mines and Resources Canada, *Petroleum Supply and Demand*, *419*.

18 The 1977 Ontario policy document by the Ministry of Energy, *Ontario's Energy Future*, *302* stated that 'there are important constraints on the increase of deliveries of natural gas from the traditional sources in Western Canada. Increases would necessitate an increase in the Trans Canada Pipe Line system, and, unless the additional supplies are of long enough duration to absorb the cost of such an increase, its economic justification will be difficult. The fact that consumption is expanding in the western provinces will reduce the quantities available for export to other Canadian provinces. Further, unless additional supplies are expected to maintain deliveries for a long period of years it will be difficult to market the additional, short term natural gas. (A new potential consumer is not likely to choose natural gas over oil unless con-

vinced that supplies will be available far into the future.) A result could be short term surpluses of natural gas in Canada, coincidental with increasing reliance upon international crude oil supplies' (p. 18).

19 See note 8 of chapter 1.

20 See W. Patterson and R. Griffin, *Fluidized Bed Energy Technology*, *315*.

21 This is true even though electrical energy capacity is projected to more than double (Figure 4), because total energy demand will also almost double. For details see Figure 2.

22 A calculation is revealing. Roughly 10^6 megawatts (MW) of installed capacity would be needed to meet total projected secondary energy demand in 1990 (see Figure 2 and Table 4). Assume hydro-electric capacity doubles by 1990 to about 73,600 MW, a more optimistic value than even Table 4 contemplates. This leaves roughly 926,400 MW to be accounted for. Assume now an annual rate of construction over the next twelve years to 1990 of six nuclear stations, each containing four 1,000 MW reactors, a construction rate well beyond present social and industrial capacity. Even so, we obtain only 288,000 MW, or about 31 per cent of the total. See also a similar calculation making the same point in *Energy Strategy*, *100*, p. 149.

The same point was made forcefully for the United States by Taylor *Energy*, *409*, *Energy Plan*, *410*; see also the Harvard Business School's Study by R. Stobaugh, et al., *Energy Future*, *398*.

23 The 'business-as-usual' approach of the Energy, Mines and Resources documents has been increasingly mirrored by publications of the Science Council of Canada. Beginning with a report that at least emphasized the importance of conservation (*Canada's Energy Opportunities*, *370*), the next report (*Canada as a Conserver Society*, *372*) made conservation a matter of urgency and added a reorientation toward renewables. However, the most recent report (*Roads to Energy Self-Reliance*, *374*), while acknowledging this policy thrust in a general way, is virtually indistinguishable from the *Energy Strategy*/LEAP orientation on which it is predicated. Conservation is demoted to the auxiliary role of rendering energy supply technology more efficient and allocated a mere $270 million – and for co-generation only – in a twenty-year research and development budget of $3.5 billion. Substantial rates of growth in energy demand are assumed and reliance is placed firmly on increased use of fossil fuels and nuclear thorium/plutonium electricity. Renewable energy technologies are allocated a trifling $136 million in the budget. The report ties increasing energy consumption to economic and social well-being and leaves the choice of technology to scientific experts while supporting the expansion of current technological commitments. See Chapter 6 and D. Cubberley's review *Science on Trial*, *71*.

24 See *Energy Strategy, 100*, p. 49
25 Response to question by P.V., Ottawa public meeting, 1977.
26 See *Toronto Star*, January 5, 1979.
27 See J.E. Gander and S.W. Belaire, LEAP, *123*, pp. 212–13.
28 A further exacerbating factor is that many of Canada's provinces will be competing with one another for available capital. In Chapters 4 and 5 we discuss Ontario's need for high levels of capital funding. Quebec will also be actively seeking foreign capital to finance electrical expansion. Evidently something like 25 per cent of all Quebec foreign borrowing in the coming years will go to such developments; see H. Lajambe *Coûts et externalites, 210* (cf. the commentary by M. Vastel, *Le Devoir*, March 14, 1978), and D. Rosenburg, *James Bay Update, 343*. The provincial debt capacity in New Brunswick is strained by the building of the Point Lepreau nuclear power station; see N. Dale, and S. Kennedy 'Toward a Safely Relevant Process of Impact Assessment,' in F. Tester, *Conference on Social Impact Assessment, 412*. These and other provincial competitions for capital can only lower optimism for the overall outcome and for Ontario especially.
29 See Science Council of Canada, *Uncertain Prospects, 373*. See also N.H. Britton and J.M. Gilmore, *Weakest Link, 26*.
30 See Energy, Mines and Resources Canada, *Financing Energy Self-reliance, 101*.
31 S. Duncan, 'Energy?', *89*.
32 See J.R. Downs, *Availability of Capital, 87*.
33 S. Duncan, 'Energy?', *89*.
34 See T. Berger, *Northern Frontier: Northern Homeland*, 2 vols. (Ottawa: Queen's Printer, 1977). See also F.J. Tester, *Socio-economic and Environmental Impacts of the Polar Gas Pipeline: Keewatin District*, 2 vols., Environmental-Social Program, Northern Pipelines (ESCON #A7-20, 21) (Ottawa: Indian and Northern Affairs, 1978). Also strongly recommended is the publication *Northern Perspectives*, published by the Canadian Arctic Resources Commission (CARC); see especially vol. 6, 1 (1978). See also CARC, *Northern Transitions*, 2 vols. (Ottawa, 1978).
35 See, for example, D. Pimlott, et al., *Oil under the Ice, 323*.
36 In the north, while energy development issues are most important, water development issues are also of crucial importance. There are already potentially large effects from the James Bay hydro-electric scheme (see Environment Canada, *James Bay Project, 108*, J.A. Spence and G.C. Spence, *Ecological Considerations, 389*, and H. Lajambe and D. Rosenberg, note 37) and the Bennett Dam on the Peace River (see the Peace-Athabasca Delta Project Group, *The Peace Athabasca Delta: A Canadian Resource*, Summary Report, Ottawa: Queen's Printer, 1972). An even larger impact is to be expected

from other proposed water diversion schemes. For a review of these latter proposals see the document *Water Diversion Proposals of North America*, published for the Canadian Council of Resource Ministers by the Alberta Department of Agriculture, Water Resources Development Planning Branch (Edmonton, 1968). For a detailed discussion of some of these proposals see R. Bocking, *Canada's Water: For Sale?* (Toronto: James, Lewis and Samuel, 1972); R. Bryan, *Much is Taken, Much Remains* (N. Scituate, Mass.: Duxbury Press, 1973); and J. Laxer *Energy Poker Game, 215* and P. Sykes *Sellout, 405*. For a particular discussion of proposals affecting northern Ontario by citizens of that region, see the publication in 1970 of *The Water Plot* by the Damn the Dams campaign (General Delivery, Thunder Bay P, Ontario). See further the Report of the Royal Commission on the Northern Environment (under Mr Justice Hartt), *Issues Report* (Toronto, 1978).

37 See W. Patterson and R. Griffin, *Fluidized Bed Energy Technology, 315*.

CHAPTER 3

1 Ontario is already beginning to feel the pinch of being a 'have-not' province in fossil fuels. In a reversal of the historical situation, Ontario is accumulating a provincial deficit, while Alberta is accumulating wealth. In 1979 the average Ontario resident paid 35 per cent more tax than the average Alberta resident for comparable social services. At present, more than $3 billion of Ontario-based capital is being invested in Alberta annually and more than 30,000 people now leave Ontario for Alberta, most of them skilled workers. See D. Hartle, *Federal-Provincial Relations, 154*, and *London Free Press*, September 25 and October 25, 1979. The further increases in the prices of fossil fuels contemplated for the period 1980–83 will exacerbate this situation.

We remind the reader here that Ontario is a 'have-not' province *only* in conventional non-renewable energy sources. The amount of available solar energy is much larger than present commercial energy consumption in the province. A rough calculation indicates that in an area of 2×10^6 square kilometres something like 3×10^{18} BTU are received annually, or about 1,000 times Ontario's current commercial energy consumption. Of course, at the present time this energy flow falls largely outside of the economic system. In the cases of passive solar space heating and agriculture, solar energy makes a major energy contribution, though in these cases it is not at present included in economic accounting. Naturally occurring wind and geothermal energy are also very substantial; see, for example, the calculation of potential wind energy in Chapter 9 and the studies cited in Energy, Mines and Resources Canada, *Renewable Energy Sources: A Guide to the Literature* (Ottawa, 1977).

Unexploited hydro-electric energy is roughly equal to currently exploited hydro-electric energy (see Chapter 9).

2 The figure of 50 per cent is some 4 per cent higher than that shown in Figure 8 and varies by 1 per cent from that in Table 6. Variations of 3 per cent or more in energy accounting are not uncommon and reflect continuing inaccuracy and ignorance. See note 21 of Chapter 7. In the specific case of oil, however, two-thirds of the difference is accounted for as demand for oil to generate electricity (see Table 6).

3 Ontario Ministry of Energy, *Ontario's Energy Future*, *302*.

4 Ontario Ministry of Energy, *Energy Security for the Eighties: A Policy for Ontario* (Toronto, September 1979).

5 See Ontario Ministry of Energy, *Ontario Energy Review*, *307*. A perusal of the demand assumptions made in this document (see pp. 41–3) will indicate that even the low energy use case still represents a substantially business-as-usual approach. By comparison see the discussion in Part III.

6 See, for example, Ontario Ministry of Energy, *Ontario's Energy Future*, *302*, pp. 52, 56, and *Ontario Energy Review*, *307*, comparing pp. 41 and 47.

7 See the discussion in Chapters 2 and 3 and in particular in *Ontario's Energy Future*, *302*, pp. 41ff.

8 See notes 2 and 5 of Chapter 2.

9 See note 22 of Chapter 2.

10 See *Ontario's Energy Future*, *302*, p. 22. We note in passing that there is a small lignite deposit, estimated at 200 million tons, at Okanawa, Ontario, sufficient to power a 1,000 megawatt electric plant for roughly thirty years.

11 The long-range forecast (LRF 48A) is summarized in Appendix B of Middleton Ass., *Alternatives to Hydro's Generation Programme*, *254*. The public document is difficult to find; a detailed plan appears in none of the Hydro documents, though an abbreviated, earlier version may be found in Table 5 of Ontario Hydro, *Load Forecasting*, submission to the Royal Commission on Electric Power Planning (Toronto, May 1976). This latter document also contains a summary discussion of successive long-range forecasting plans that included a plan for still higher rates of growth in electrical generating capacity; see also the discussion in notes 5 through 8 of Chapter 3 of the Royal Commission on Electric Power Planning, *A Race Against Time*, *354*, and their Figure S-13, p. 32 and accompanying text; cf. our Figure 10. For the 1980 statement on the expansion of Ontario's electrical generating system, see Ontario Hydro's *1980 Review, System Expansion Program* (Toronto, March 1980).

12 Table 6 shows that nuclear electricity accounted for about 15 per cent of total 1974 electrical energy demand. In that year there was only about 2,000 MW of installed nuclear capacity out of a total of about 18,000 MW of installed electrical generation capacity. In 1979 nuclear capacity amounted to 5244

megawatts out of a total of 23,698 megawatts of dependable capacity. According to Hydro projections in 1980, there will be a further 9612 megawatts of
nuclear capacity and 966 megawatts of coal capacity. giving nuclear capacity
only about 44 per cent of total installed capacity, But over the very brief experience Ontario Hydro has had with nuclear plants they have been able to run,
and been made to run as a matter of economic policy, at a higher load factor
than fossil-fuel plants – roughly 80 against 60 per cent – and this will bring
expected nuclear energy supply up to about 57 per cent of total electrical
energy supplies.

13 See *Energy Security for the Eighties* (Toronto 1979), p. 14; and Ontario Hydro,
     1980 Review: System Expansion Program (Toronto 1980), p. 14–2.
14 See Ontario TEIGA *Submission*, *310*, p. 33.
15 Ibid., p. 32.
16 It is noteworthy that the Select Committee Enquiring into Hydro's Proposed
     Bulk Power Rates, which was established in order to find ways of reducing
     Ontario Hydro rate increases for 1976, only managed to shave 5 percentage
     points from Hydro's effective application for a 31 per cent increase. For 1977
     the Select Committee recommended an annual increase of 34 per cent, though
     this was subsequently reduced to 25.6 per cent. For 1978 and 1979, when the
     impact of the general economic slowdown was clearer, effective rate increases
     of about 5 per cent and 3 per cent respectively were granted, but the proposed
     rate increase for 1980 has jumped to 16 per cent.
17 See Ontario Hydro, *Socio-economic Factors*, *294*.
18 See Ontario Hydro, *Review of the System Expansion and Financial Plans*,
     submission to the Royal Commission on Electric Power Planning, Toronto:
     April 1977, p. 35.
19 See ibid., and Ontario Hydro, *Socio-economic Factors*, *294*.
20 Ontario Ministry of Energy, *Ontario's Energy Future*, *302*, p. 51.
21 See, for example, Ontario Ministry of Energy, *Ontario's Energy Future*, *302*,
     pp. 3–4, 23–4, 45–6, and the discussion in Chapter 5.
22 See Ontario Ministry of Energy, *Ontario's Energy Future*, *302*, p. 51.
23 Ibid., pp. 23, 24.
24 Ontario Hydro, *Energy Security for the Eighties*, pp. 11, 14.

CHAPTER 4

1 *Ontario's Energy Future*, 302, esp. pp. 10, 20, 23–4, 45–6. See also R.M.
    Dillon, *Ontario-Nuclear Electric*, *84*; J. Laxer, *Canada's Energy Crisis*, *216*,
    pp. 59–60 on federal commitments. Again we remind the reader that discus-
    sion of the potential contributions of electrical energy to total demand, espe-
    cially prior to 1977, should be considered carefully in the light of the adopted

conversion rate from kilowatt-hours to BTU; otherwise an artificially inflated electrical energy contribution can easily result.

2 This has been the thrust of a recent argument by the Canadian Nuclear Association.

3 Energy, Mines and Resources Canada, *Energy Strategy*, *100*, pp. 68–70, 143–5. For example, in discussing its projections of future installed electrical capacity, the document states: 'In this connection it would appear prudent, to some degree, to base capacity expansion plans at least initially on projected load growths that are closer to the upper end [the high demand end] of the plausible range' (p. 69). The nuclear component of the projection suggests that nuclear capacity might approximate 22,000 megawatts to 1990 and account for about 18 per cent of total national electrical capacity by that time, though only 57 per cent of Ontario's capacity (see note 12 of Chapter 3). This attitude, which clearly commits the country to advanced fuel cycle technology, might be contrasted to the caution in the Ford Foundation's Report of the Nuclear Energy Policy Study Group *Nuclear Power*, *280*, which concluded by cautioning the United States against any early commitment, even on a demonstration basis, to reprocessing of spent fuel. AECL's past enthusiasm for nuclear power has not gone unnoticed or unopposed. See 'AECL Stresses Need to Develop Plutonium,' *Globe and Mail*, May 16, 1977; and *Globe and Mail*, June 24, 1977. For the historical background to present nuclear power commitments, see G.B. Doern, *Atomic Energy Control Board*, *85*; W. Eggleston, *Canada's Nuclear Story*, *96*; J.E. Hodgetts, *Administering the Atom*, *170*; I. Hornby and B. McMullan, *The Nuke Book*, *181*; F. Knelman, *Nuclear Energy*, *204*; G.C. Lawrence, *Science Forum*, *54*; P. Metzger, *The Atomic Establishment*, *251*; and D. Torgerson, *Dream to Nightmare*, *420*.

4 See Energy, Mines and Resources Canada, *Management of Canada's Nuclear Wastes*, *102*.

5 It is interesting to observe the multiple interconnection in jobs held by a small group in the nuclear energy field. For example, in 1978 G.M. MacNabb simultaneously held the positions of deputy minister of Energy, Mines and Resources, Canada; president of Uranium Canada, the federal holding company for uranium reserves; and member, Board of Directors of AECL. Such multiple responsibilities may well give rise to conflicts of interest and responsibility; in any event they do evidence the rather closed manner in which the Canadian nuclear commitment has developed historically and in which it still operates today. See Nichols' study for the European Office of Economic Cooperation and Development, *Public Participation in Energy Policy in Canada*, note 25, Chapter 9, and the Interim Report of the Royal Commission on Electric Power Planning, *A Race Against Time*, *354*.

6 See Ranger Uranium Environmental Enquiry, *First Report*, *331*; Royal
   Commission on Environmental Pollution, *Sixth Report*, *356*; and Royal
   Commission on Electric Power Planning, *A Race Against Time*, *354* for
   detailed examinations. Specifically on U.S. plutonium security see J. McPhee,
   The Curve of Binding Energy, *248*; and the Canadian Coalition for Nuclear
   Responsibility, *Nuclear Power*, *40*. On the plutonium economy generally see
   also notes and 3 and 15 (this chapter), and the report of the National Council
   of Churches, *The Plutonium Economy: A Statement of Concern*, with appendices
   (available from 475 Riverside Drive, New York, NY 10027).
7 See Royal Commission on Electric Power Planning, *A Race Against Time*,
   354. The report called attention to the difficulties in obtaining capital and
   manpower for rapid expansion of nuclear electric generating technology. It
   documented in some detail both the uncertain future for uranium supplies
   and costs and the uncertain economic future of the Canadian nuclear industry,
   which is unlikely to survive in its present form unless there is rapid expansion
   in orders for nuclear power plants. The report concluded, as we do, that the
   nuclear electric future would likely be neither cheap nor secure.
      The report also provides an excellent summary of the environmental and
   health hazards associated with nuclear power. It emphasizes the importance
   and ineliminability of the human error factor in nuclear accidents, and it supports a moratorium on expansion of nuclear capacity beyond 20,000 megawatts (our ceiling would be much lower) until it has been 'demonstrated
   beyond reasonable doubt that a method exists to ensure the safe containment
   of long-lived, highly radioactive waste for the indefinite future.' Beyond this
   two important hazards associated with uranium mining are emphasized: the
   contamination of Ontario lake waters by radioactive wastes leached from
   uranium mine tailings; and the difficulties surrounding successfully 'decommissioning' uranium mines. In 1979 there were more than 75 million tons of
   mine tailings in the Elliot Lake region and 'several lakes and streams [had]
   been badly contaminated.' With the expansion of the mining industry associated with an expanded nuclear programme, 'it is foreseeable that several
   hundred million tonnes of mill tailings will be left behind,' which will 'constitute an increasing health and environmental problem.' The decommissioning
   problem centres around the disposal of these tailings when there is no further
   economic incentive to take an interest in the area. The report concedes that
   although there are 'legislative proposals ... to finance the mine decommissioning phase, there are no obvious technological solutions in sight' (p. 71).
      The report is perhaps most critical of the social and political dimensions of
   the nuclear field. It stresses the likely political consequences of a complex,
   highly centralized, potentially dangerous technology and reviews with concern

the evidence on international nuclear weapons proliferation and domestic nuclear blackmail. Throughout, the report is critical of Canadian governments for lack of responsible leadership concerning nuclear hazards and of both government and nuclear power agencies – notably Ontario Hydro and Atomic Energy of Canada Limited – for not encouraging, or even thwarting, the provision of public information and effective public involvement in decision-making (cf. note 32 below).

The weakest part of the report revolves around its ambivalent attitude to the social and political consequences of nuclear power. On the one hand, the report documents the social and political hazards associated with expanding nuclear power; on the other hand, it is loathe to look elsewhere for future energy supplies. The result is an avoidance of the issues through vagueness. For example, the report's review of the social, ethical, and political issues is the briefest chapter in the report; yet from our point of view it concerns one of the most crucial areas. The report concludes that while implementing existing CANDU technology does not threaten civil liberties, advanced fuel cycle technology might well do so, but does 'not endorse such an initiative.' With respect to existing technology, not seven pages further on the report argues: 'we have concluded that a comprehensive crisis management system incorporating police and political authorities should be designed and available to deal with the complex and rapid judgements and decisions which a nuclear incident might make necessary. The demonstrated existence of such a response capability might prove to be a considerable deterrent to threats or incidents involving Ontario's nuclear programme' (p. 162).

Thus the report concedes that nuclear power creates an unprecedented requirement to put in place measures for public control otherwise seen only in war. With respect to the admittedly greater threat posed by advanced fuel cycle technology, the report's previous chapters have argued eloquently the case we also have argued that economic and social investments in nuclear technology under the existing policy commitments will be so great by 2000 that society will be virtually irrevocably committed to advanced fuel cycle technology. The report continues with a characterization of the legitimate role, and cultural importance, of minority groups within society which is, in our view, far too weak to do justice to the notion of a democracy. We go on to explore some of these social, ethical, and political issues in the next chapter and in Part IV.

8 See for examples the references of note 6 above and the RG, F.4.
9 Again an example is instructive. Suppose that the total energy requirements in Ontario were to double between 1980 and 2000, representing a modest growth rate of about 3.5 per cent annually, and only one-half of that energy were to

be supplied by 2,000 megawatt Pickering type nuclear power plants. Then approximately fifty-five power plants would be required by 2000, representing 110,000 MW capacity for a total investment of something in excess of $120 billion at a rate of construction of over two and a half Pickering-sized plants a year between 1978 and 2000.

10 Ontario Ministry of Energy, *Ontario's Energy Future, 302*, pp. 23–4. Note that these assertions entirely neglect solar and biomass energy, principally among renewable energy forms, both of which satisfy all of the constraints mentioned. Moreover, biomass energy represents a comparable energy source to current provincial demands and the flow of solar radiation is many hundreds of times that of current provincial demand.

11 *Nuclear Engineering International* (September 1974), p. 741. Much the same considerations apply to key fossil fuels; hence the recent diversification of the big energy companies; see J. Ridgeway *The Last Play, 333*.

12 Fissile uranium-235, an isotope of the common uranium-238, occurs naturally as about 0.7 per cent by weight of uranium oxide ore at best. Typically, though not for CANDU reactors, this ore is 'enriched,' in a difficult and expensive process, to create a reactor fuel containing between 2.5 per cent and 3.5 per cent uranium-235. Even then, not all of the fissionable uranium can be exploited before various reactor processes force the removal of the fuel. For a convenient discussion see C. Law and K. Glen, *Critical Choice, 214*.

13 Ibid.; cf. note 14 below.

14 For a discussion of the Canadian situation, see the recent federal study by L.K. Hare, *Management of Nuclear Wastes, 102*. This report takes a fairly optimistic view of the matter in comparison with many other studies; for example, the Report of the British Royal Commission, *Sixth Report, 356*, as well as W. Rowland, *Fueling Canada's Future, 353*, and R.J. Uffen, *Disposal of Used Fuel, 425*. A comparison of the Hare and Uffen reports is contained in the Interim Report of the Royal Commission on Electric Power Planning, *A Race Against Time, 354*, Chapter 6.

15 On the first incident see the discussion in the *London Free Press*, September 24, 1975. The second incident occurred on April 2, 1978; see *Globe and Mail*, April 14, 1978. The third incident is reviewed in M.C. Olson, *Unacceptable Risk, 287*, pp. 32–4. The first of the reactor core incidents – that concerning the Fermi reactor in Chicago – is examined in detail by J.G. Fuller *We Almost Lost Detroit, 122*. The second accident, that at Three Mile Island, March 28 to April 8, 1979, was reported in every major newspaper, see, for example, *New York Times* for the period. A special issue of *Spectrum* (Journal of the Institute of Electrical and Electronic Engineers Ltd.), November 1979, is devoted to an analysis of the event and the problem, also C. Cina and T. Goldfarb, *Three*

Mile Island, 50. The matter was investigated by a presidential committee which was harshly critical of nuclear safety procedures and recommended sweeping changes in the U.S. nuclear regulatory structure; see *London Free Press*, October 31, 1979. Other investigations are still proceeding. For a different, highly unlikely, nuclear-related accident, which nonetheless actually occurred, see *Diesendorf*, note 27, Chapter 9.

On like questions in Canada see especially Canadian Coalition for Nuclear Responsibility, *Nuclear Power*, *40*, F. Knelman, *Nuclear Energy*, *204*, R. Torrie, *Reactor Safety*, *422*, *People*, *421*; see also the letter by 'M. Schulz' later revealed as W. Taves, a control room operator at Bruce Nuclear Station until May 1979, to Mr. E. Sargent, Liberal MPP, April 16, 1979. The full text of the letter appears in *Birch Bark Alliance* 2, 2 (1979), which also supplies an analysis of the Three Mile Island incident (*Alliance* is available from OPIRG, Trent University, Peterborough K9J 7BH). Earlier, a shift foreman at a Hydro nuclear plant, K. Millray, resigned for similar reasons; see *Globe and Mail*, May 16, 1979. As recently as May 1979 an accident at Hydro's Bruce B nuclear plant exposed two workers to 'high levels' of radiation and disclosed new weaknesses in the safety system; *Toronto Star*, May 5, 1979. The attitude of such critics as H. Caldicott, *Nuclear Madness*, *38*; F. Knelman, *Nuclear Energy*, *204*; D. Hayes, *Nuclear Power*, *156*; and W. Patterson, *Nuclear Power*, *313*, should be contrasted to the attitudes of proponents of nuclear power such as C. Law and R. Glen, *Critical Choice*, *214*, who do not mention any of these incidents directly. On the dissemination of nuclear information see notes 17 and 32 below.

16 See Nuclear Regulatory Committee, *Reactor Safety Study*, *281*, for the U.S. study used as basis for Canadian claims. The figure in the text was given by Mr Morison of Ontario Hydro in testimony before the Royal Commission on Electric Power Planning, see Transcript, vol. 206, p. 32807. Cf. the critical discussion of safety figures by R. Torrie and G. Edwards, *Summary Argument*, *423*. See also note 27 of Chapter 9.

17 The most famous of these reports at the present time is that of the team headed by Rasmussen, U.S. Nuclear Regulatory Commission, *Reactor Safety Study*, *281*. This report has been heavily criticized in many places. See Union of Concerned Scientists, *Risk of Nuclear*, *428*, and T.P. Speed, *Negligible Probabilities*, *388*. (More generally, see the account by Salzman in *Index on Censorship* (1958), 39; by G. Hardin, 'The Fallibility Factor,' *Skeptic*, 14 (1976), 10; and *Bulletin of the Atomic Scientists* (November 1976), 25; see also Tversky/Kahneman, 'Judgement under Uncertainty,' *Science* 185 (September 1974). For a critique of the application of the Rasmussen report to the CANDU

system, see R. Torrie, *Reactor Safety*, *422*. Besides engineering calculations and related factors, the entire conception of reaching a rational decision about acceptable risk employing the usual economic technique of cost/benefit analysis has been seriously disputed. See RG, F.4.

18 Numerous studies have been done on the economics of nuclear power, with wide divergences among them. Of principal interest here is that by Ontario Hydro on which we base our discussion; a subsequent report by S. Bannerjee and L. Waverman, *Life Cycle Costs*, *9*, which is slightly more favourable to nuclear power; the opposing study by A. Lovins *Soft Energy Paths*, *231* (cf. the earlier study *Nuclear Futures*, *227*); the supporting study by Komanoff Associates *Comparison of Nuclear and Coal*, *207* (cf. Komanoff *Doing without Nuclear*, *208*, L.M. Berger *Nuclear Power*, 18); and the Interim Report of the Royal Commission on Electric Power Planning *A Race against Time*, *354*, which follows the Ontario Hydro Study *Comparative Costs*, *296* and that by H. Swain, et al. *Canadian Prospects*, *402*. We could spill a lot of ink comparing these studies. There are, however, several general lessons we wish to draw from them. First, it all depends on what are counted as costs. Counting differently from the nuclear industry, Friends of the Earth arrived at a nuclear-electric cost figure five times higher than the conventional figure. J. Harding, *The Deflation of Rancho Seco 2* (San Francisco: Friends of the Earth, 1975) and I.C. Bupp, J.C. Darrien, et al. *Economics of Nuclear*, *36*, point out that the capital cost difference between nuclear and coal plants has been increasing annually for some time; Bannerjee/Wavermann, for example, do not mention such effects. Second, the prices of most of the key items are highly politically sensitive and are estimated differently by those who estimate the politics differently, including those who ignore politics. See our discussion of uranium prices, notes 22, 25 below and *New Scientist* (June 26, 1975) on the costs of reprocessed fuel. Similarly, the costs of oil and gas are now widely acknowledged to be 'policy priced.'

The point is that, taken together, there is enormous uncertainty in the costs of energy. We have simply illustrated those uncertainties and argued that they obscure any claims that short-to-medium term cost comparisons give a decisive advantage to any one energy technology. In the long term it is intuitively reasonable to view the renewable technologies as having decisive advantages, since they combine simple, more stable technologies with long-lived fixed investment in design, insulation, and urban form.

19 Ontario Hydro, *Comparison of Costs*, *296*. But see Ontario Hydro Report 58431 (January 1979).

20 See, for example, A.B. Lovins, *Soft Energy Paths*, *231* and *Electric Utility Investments*, *234*.

21 As an illustration, we cite the additional billion dollars that was subsequently added to cost estimates for the two Bruce heavy water plants (*Globe and Mail*, July 12, 1978) and the additional cost of a million dollars, excluding costs paid by the manufacturer, for boiler repairs at the Bruce plant that may yet run well above that figure; see *Globe and Mail*, August 15, 1978, and *London Free Press*, November 1, 1979.

22 See Ontario Legislature, Select Committee, *Staff Summary, 300*. The uncertainties we see for the nuclear field, in particular for fuel costs and availability, the viability of the Canadian nuclear manufacturing industry, and nuclear manpower difficulties, are all supported in the Interim Report of the Royal Commission on Electric Power Planning, *A Race against Time, 354*, pp. 139, 147–9, 179, 135, 145.

23 D.S. Robertson and Associates, *Brief Review of Uranium, 336*; see also *Staff Summary, 300* and note 21 above.

24 See Ontario Hydro, *Submission to the Select Committee, 290*.

25 This margin was estimated by comparing the forecast costs of the uranium fuel element as given in Ontario Hydro's *Economic Forecasting Series* (June 1975) and the equivalent costs escalated at Ontario Hydro's forecast of the world price of uranium as submitted to the Select Committee of the Legislative Assembly Enquiring into Hydro's Proposed Bulk Power Rates, June 1976.

26 See Dennison Mines Ltd, *The 'Uranium Cartel Myth', 80*. All quotations are from this document. The opinion that normal marketing arrangements have been far exceeded was supported by expert testimony to the Select Committee.

27 See S.S. Penner and L. Icerman, *Energy, 317*; B.W. Erickson and L. Waverman, *Energy Question, 112*; L.C. Ruedicili and M.W. Firebaugh, *Perspectives on Energy, 357*; A.R. Flower, 'World Oil Production,' *114*; C. Watkins and M. Walker, *Oil, 445*; R.H. Williams, *Toward a Solar Civilization, 457*; and C.L. Wilson, *Energy: Global Prospects, 459*.

28 Nationalization of uranium mining companies was explicitly rejected by the Ontario Government. See the statement by James Taylor, Minister of Energy, to the Select Committee of the Legislature, January 11, 1978. In contrast to the situation with respect to uranium, the North American and world supplies of coal are so large and geographically diverse that coal supplies are likely to remain competitive. See ibid. Even the federal energy strategy document projects an almost fourfold increase in the availability of Canadian coal over the next fifteen years and the creation of an export surplus. Moreover, this increase is expected to be obtained comparatively cheaply whether measured in absolute terms (the $3.2 billion, 1975 dollars, of capital investment required), or in $/BTU. In addition, it is unlikely, though no doubt possible, that U.S. supplies of coal in Ontario could be curtailed for reasons other than a

real shortage of supplies. Such a move could be costly to the U.S. coal producers and an unnecessarily heavy handed means of exercising control over an already dependent economy.

29 The fact is that the state of medical science does not yet permit a very precise estimate of increased risks from exposures of this sort; see W. Jacobi *Intake of Radon Daughters, 195* and notes 7 and 8 of Chapter 5. For accounts of the episode see reports in *Globe and Mail*, July 30, 31, 1975; see also related incidents reported in the *Globe and Mail* on July 21 and August 6, 1975.

30 We have been fascinated to learn that a large number of people in Port Hope subsequently signed a petition in 1978 requesting that a proposed new plant for spent nuclear fuel also be located near Port Hope. We understand that their overwhelming fear is that a different site for the new plant may mean closure of the old plant with a subsequent loss of jobs. Thus can a centralized system force a choice between economic well-being and freedom from specific risks that its own enterprise generates.

We have been equally fascinated to learn that AECL has apparently found the original dump site too costly to clean up and move and has in fact had the license for the site extended; see G.B. Knights, Atomic Energy Control Board Memorandum 15-2-E1, July 6, 1975. Though the new use conditions are undoubtedly more stringent, it is the inherent economic and bureaucratic inertia in the system that we mark. See in these respects the repair of Brown's Ferry Plant (M. Olsen *Unacceptable Risk, 287*, p. 34), especially vis-à-vis the history of contradictory reportage and repair difficulties connected with the Three Mile Island plant as presented in *Spectrum* (November, 1979) and C. Cina and T. Goldfarb, *Three Mile Island, 50*.

31 The Canadian Nuclear Association has already made clear its view that to maintain a viable nuclear industry Canada should pursue a vigorous policy of reactor sales abroad. This is roughly because they do not expect domestic sales to be of sufficient quantity. In any case, the Association has also emphasized that the nuclear industries will push hard to see that they function at full capacity and that they continue to expand.

32 Two recent royal commissions have stressed the awesome potential for the restriction of public freedom in the case of nuclear power, see Ranger, *First Report, 331* and Flowers, *Sixth Report, 356*. See also S. Ebbin and R. Kasper, *Citizen Groups, 93*, D.P. Gesaman and D.E. Abrahamson, *Dilemma of Fission Power, 125*, J.G. Speth et al. *Plutonium Recycle, 390*, A. Lovins, *Soft Energy Paths, 231*. For aspects of the Canadian situation see the Interim Report of the Royal Commission on Electric Power Planning, *A Race against Time, 354*, with its running theme of insufficent public information, pp. 65, 73, 96–7, 101 and 155, and the review by Lydia Dotto, *Science Forum* (November/December 1978), p. 41.

33 We note cautiously that there seems to be growing provincial government support for this point of view. A September 1979 position paper by the Minister of Energy (*Energy Security for the Eighties: A Policy for Ontario*) for the first time gives equal support to renewable and nuclear development, while remaining silent on the question of further expansion of the nuclear system beyond present commitments (*Globe and Mail*, October 2, 1979). However, the political dimensions to energy policy are such that one cannot deduce firm policy commitments from this kind of document.

CHAPTER 5

1 See RG, H. For detailed studies of the Canadian situation see Bureau of Municipal Research, *What Can Municipalities Do About Energy? 37*; G. Desfor and M. Miloff, *Urban Form and Energy Consumption, 81*; J. Hix, *Energy Demands, 169*; Middleton Associates, *Energy Management at the Local Level, 255*; D.B. Brooks, *Zero Energy Growth, 28*; W. Leiss, *Quality of Life and Energy Conservation, 222*. Special thanks are due to Michael Miloff, Middleton Associates, for assistance in this area.
2 B. Hannon, *Energy, Labour and the Conserver Society, 146*.
3 One should note that if decentralized energy technologies are in fact largely taken over by giant centralized corporations, many of their potential employment and other advantages may disappear. For discussions of the impact of different energy technologies on employment kinds and levels see RG L.5. and note 1 above and notes 18 of Chapter 7 and 37 of Chapter 8.
4 Ontario Ministry TEIGA, *Design for Development, 309*.
5 Note the historical relationship between attempts to centralize the technology and attempts to gain control of the industry. See, for example, the discussions by J. Ridgeway and B. Connor *New Energy, 334* and especially that by S. Novick, *The Electric Power Industry, 278*; see also the discussion of Ontario Hydro in H.V. Nelles *The Politics of Development, 272*, in W.T. Easterbrooke and H.G. Aitken, *Canadian Economic History, 92*, P.C. Hill *The Social Costs of Electric Power, 167* and J.O. Dean *Role of Ontario Hydro, 78*.
6 Ontario Ministry of Energy, *Ontario's Energy Future, 302*, pp. 40–9.
7 See the recommendations of the International Commission on Radiological Protection, *Publication 9, 193* and *Publication 15, 194*. The present standards are 5 rems per year, up to 3 in a single quarter, for radiation workers and 0.17 rem per year for the general public, excluding natural and medical exposure in both cases. Further, see the discussion in W. Jacobi, *Intake of Radon Daughters, 195*; and in J. Sevc, et al., *Lung Cancer, 378*. The situation is reviewed in G. Edwards, *Estimating Lung Cancers, 94* and *Nuclear Wastes, 95*. Most recently see

the serious results presented by T.F. Mancuso, et al., *Radiation Exposures*, *241*, J.L. Marx, *Low-Level Radiation*, *243*, and C. Holden, *Low-Level Radiation*, *171*. Such considerations have recently led the international authority K.Z. Morgan to suggest substantial further reductions in the admissible exposure levels (see *Cancer*, *258*), and other cancer researchers to speak out increasingly forcefully on the issues (see, e.g., the statement by the well-known authority S. Epstein, *Politics of Cancer*, *111*).

In fact, what has happened over the past thirty years is that medical opinion has slowly shifted from an unconcern with low-level radiation (an assumed 'threshold' effect) to an admission that damage may be proportional to dosage (a linear relation) to the present increasing concern with the possible disproportionate impact of low-level radiation (a faster-than-linear relation). What of the future? Public policy, however, shifts more slowly; see the conservative stances of recent U.S. reports: U.S. National Academy of Sciences, BEIR *Report 3*, *269*, and U.S. H.E.W. *Ionizing Radiation Report*, *433*.

8 See the extended discussion in the *Final Report* of the Royal Commission on the Health and Safety of Workers in Mines (J.M. Ham, Commissioner) Toronto: 1976. See also A.S. Schurgin and T.C. Hollacher *Radiation Induced Lung Cancers*, *369* and United Nations *Radiation*, *429*. For one account of the testimony discussed in Saskatchewan uranium mining history see W. Harding *Nukenomics*, *150*, *Uranium*, *151*; see also Saskatchewan Legislature, *Report*, *362*. It is noteworthy that in late 1977 the Ontario Workmen's Compensation Board accepted its first recognized case for legitimate compensation from cancer caused by low-level radiation; see *Globe and Mail*, December 29, 1977.

Much more common are the health problems caused by silicone dust, coal dust, and other airborne particulates. This problem has not at all been confined to uranium mining; it is, and remains, a severe problem in any proposed increase in the use of coal as an alternative energy fuel. But the battle for mining safety has been particularly bitter in the case of uranium mining. On these issues in general see, for example, S. Epstein, *Politics of Cancer*, *111*, L. Tartaryn *Dying for a Living*, *408*, and the testimony discussed in the *London Free Press*, May 29, 1975.

9 See R. Gillette, 'Transient Nuclear Workers: A Special Case for Standards,' *Science* 186 (1974), pp. 125–9; B.J. Verna, *Radioactive Maintenance*, *437* and 'Radiation Exposure and Protection Problems in Nuclear Power Plants: United States Overview,' *Atom Kernen*, 30, 4 (1978), pp. 282–6.

10 See, International Commission on Radiological Protection, *Publication 9*, *193*, *Publication 15*, *194*.

11 H.G. Ibser, *Genetic Roulette*, *185*.

12 See notes 15 and 17 of Chapter 4.

13 H. Inhaber, *Risk with Energy, 187* and *Risk of Energy Production, 188.* Inhaber's critics include especially John Holdren, a major anti-nuclear researcher widely referred to by Inhaber; see the text of Holdren's letter to Dr. A.T. Prince, president Atomic Energy Control Board of Canada, which says in part: '[The report is] shockingly incompetent and inaccurate ... in every section I have examined in detail ... I have found not only misreadings and misrepresentations ... but also double counting, arbitrary upward "correction" factors, and glaring calculational errors.' The full text of the letter is reprinted in *Canadian Renewable Energy News* (April 1979).

14 R.M. Dillon, *Ontario-Nuclear Electric?*, 84

15 Ontario Ministry of Energy, *Ontario's Energy Future, 302*, p. 3.

16 Ibid., p. 28.

17 See A. Weinberg, *Social Institutions and Nuclear Power, 449*, p. 27; G. Hardin, *Faustian Bargain, 148*; A.V. Kneese, *Faustian Bargain, 202*, among many others; see also note 17 of Chapter 4. Compare in this respect Veblen's earlier faith in technology as expressed in his *The Engineers and the Price System, 1921* (Viking Press, 1933), pp. 135–7. See also the comments by Simmons in H.S.D. Cole, et al., *Thinkng about the Future, 54*. To be fair, Weinberg himself is well aware of the social dimensions to nuclear power; for example, see *Social Implications of Nuclear Energy, 448* and *Dimensions of Scientific Responsibility, 450*. However, consider Weinberg's position against the wider critique of 'Systems-Engineering' projections of the future, for example as found in Cole; D. Dixon, *Alternative Technology, 82*; and R. Theobald, *Habit and Habitat, 415*, and Theobald and Scott, *TEGS 1994, 416*. It is now becoming increasingly widely recognized that the 'hard' high energy path (nuclear, fossil) is very likely to reduce drastically future flexibility of energy choice; see, for example, D.J. Crossley, *Energy Policy in Australia, 69*; L.N. Lindberg, et al., *Energy Syndrome, 224*; M. Lonnroth, et al., *Energy in Transition, 22*.

CHAPTER 6

1 These components coincide with the qualitative priorities that have also emerged in other recent studies; in particular they are precisely the options emphasized by a careful 1979 Harvard Business School study for the U.S. economy; see R. Stobaugh, G. Yurgin, et al., *Energy Future, 398*.

The reader will find two dichotomies in the literature on 'alternative' energy strategies: renewable/non-renewable and soft/hard energy technologies. It is as well to be aware that these are not the same. We use the first in this part and take it to distinguish energy technologies based on solar, gravitational, and geothermal energies ('lifetimes' of the order of billions of years) from all

others. We eschew the confusing use found in various government documents
(e.g., *Ontario's Energy Future*, *302*) which includes both plutonium/thorium
fission and fusion in the renewable category. We take soft technologies to be
those with low negative environmental and social effects, low-risk profiles that
are open to democratic control and permit local variety, experimentation, and
so on. It is important to remember that while it is a fact that all non-
renewable technologies are hard, it is not true that all renewable technologies
can be expected to be soft: consider, for example, mammoth hydro-electric
projects or photovoltaic energy 'farms.' Indeed, developed in a centralized,
'high technology' form, all renewable energy technologies could turn out to be
hard. See the discussion in H. Swain et al. *Canadian Renewable Energy
Prospects*, *402*, in C.A. Hooker and R. van Hulst, *Institutions and Energy
Policy*, *173* and note 3 of Chapter 5.

CHAPTER 7

1 Energy, Mines and Resources Canada, *Energy Strategy*, *100*, pp. 92–4.
2 See K. Linton, *Energy Conservation*, *225*; A.B. Lovins, *Energy Efficient
  Futures*, *229*. See also note 1 of Chapter 5.
3 See, for example, G.M. MacNabb, *Canadian Energy Situation*, *238*, p. 13.
4 See Energy, Mines and Resources Canada, *Energy Conservation in Canada:
  Programs and Prospectives* (Ottawa, 1977).
5 G.M. MacNabb, *Canadian Energy Situation*, *238*.
6 To remind the reader of the primary/secondary energy distinction, discussed
  in Chapter 1, primary energy demand is total energy entering the technological energy-producing system, while secondary energy is direct end-use energy
  or output from the technological energy-producing system. The difference
  between the two is simultaneously a measure of the efficiency of the energy
  production technologies and of the amount of energy dumped as 'waste'
  during the production of demanded energy. The report by D.B. Brooks et al.,
  Scenarios of Energy Demand, *30* contains a discussion of the primary/
  secondary distinction; its policy implications are reviewed in D.B. Brooks,
  Conservation, *29*.
7 D.B. Brooks, *Conservation*, *29*, p. 51.
8 A.B. Lovins, *Energy Efficient Futures*, *229*.
9 D.B. Brooks, *Conservation*, *29*, pp. 53–4. Brooks goes on to suggest that there
  are serious limits to wealth associated with these goals, but this seems
  plausible only if one assumes continued inequities in the consumption of
  material commodities and downplays non-material forms of wealth (for
  example, services, recreation, culture).
10 For studies of energy conservation cited in the text, and others, see RG, G,
   H.1, L.3. On the relation of energy to economic growth see, for example,

M.H. Ross and R.H. Williams, *Energy and Economic Growth*, *346*. U.S. Council on Environmental Quality, *Good News about Energy*, *431*. See also note 1 of Chapter 5. Industry studies come to similar conclusions; see for example, Royal Dutch Shell as reported in the *Financial Times*, June 22, 1979.

11 See CONAES, *Energy Demand*, *56*; H. Brooks and J.M. Hollander, *Energy Alternatives*, *31*; U.S. National Academy of Sciences, *Report 270*; M. Whiting, *Industry Saves Energy*, *453*.

12 U.S. Council on Environmental Quality, *Good News about Energy*, *431*.

13 Ford Foundation, *A Time to Choose*, *118*; *Exploring Energy Choices*, *117*.

14 Ford Foundation, *A Time To Choose*, *118*.

15 G.N. Hatsopoulos, et al., *Capital Investment to Save Energy*, *155*. See also T. Alexander, *Industry*, *1*; J. Darmstadter, et al., *How Industrial Societies Use Energy*, *76*, and *International Variations in Energy Use*, *77*; L. Schipper, *Productivity of Energy Utilization*, *364*; L. Schipper and A.J. Lichtenberg, *Efficient Energy Use*, *365*; T.F. Widmer and E.P. Gyftopoulos, *Energy Conservation*, *454*.

16 L. Schipper and J. Darmstadter, *Energy Conservation*, *366*.

17 D. Chapman, *Taxation and Solar Energy*, *46*; *Prepared Statement*, *45*.

18 B. Hannon, *Energy Conservation and the Consumer*, *144*; *Energy, Labour and the Conserver Society*, *146*; *Prepared Statement*, *147*; *Energy, Growth and Altruism*, *145*.

19 Ontario Ministry of Energy, *Submission*, *301*, p. 2–1.

20 Ontario's investments to stimulate energy conservation were $1.9 million in 1975–76; $2 million in 1976–77; $4.6 million in 1977–78; $5.35 million in 1978–79; and $7.187 million in 1979–80. See Ministry of Energy, *Printed Estimates, 1979–80* (Toronto, 1979). These figures should be compared with the even lower investment in stimulating development of renewable energy technologies: $1.12 million in 1977–78; $2.455 million in 1978–79; and $2.59 million in 1979–80.

The quoted conservation goal in the text is from Ontario Ministry of Energy, *Ontario's Energy Future*, *302* (p. 31) and represents a reduction in growth of less than 2 per cent per annum, something modest price rises or the present economic slowdown could in themselves achieve. Even so, 2 per cent of the provincial energy budget is still twenty to fifty times the investment in stimulating conservation.

As remarked, considering the limited resources available to it, the Energy Management Programme appears to have been quite successful. In one year, the colleges and universities of the province managed to shave $4.2 million off their energy bills. The Ministry of Industry and Tourism Energy Bus, in operation since September 1975, visited 252 firms throughout the province up to September 1976 and identified total potential savings of $15.7 million.

21 Whereas Ontario Hydro indicates roughly 6 per cent of electricity is used for space heating in the commercial sector (see *Energy Utilization and Electricity, 291*, Table 6.4-1), V.R. Puttagunta, *Temperature Distribution, 328* indicates roughly 51 per cent of all electrical demands in the commercial sector goes to space heating and air conditioning. Both sources agree that electricity represents about 20 per cent of the total commercial secondary energy demand. This, however, conflicts with the Ontario Ministry of Energy's figure of 35 per cent (per *Energy Review, 307*, p. 46). Perhaps these differences are reconcilable by different patterns of electrical end-use demands in different provinces of Canada, in which case there is much greater scope for reduction of electrical demand through substitution of solar-derived space heating in other provinces than in Ontario. In any case they reflect sadly on the confusing state of energy reportage in Canada. Cf. note 2 of Chapter 3.

22 See, for example, Ontario Hydro, *Energy Utilization and Electricity, 291*; C. Conway, *Soft Energy Paths, 62*; and *Conservation, 64*.

23 See note 22 above. Improvements in the efficiency of electric motors are now advancing rapidly – for example, see *Maclean's* October 15, 1979, for a new variable-drive for electric motors that will substantially improve efficiency.

24 For details, see E.P. Gyftoupoulos et al., *Potential Fuel Effectiveness, 139*, A.P. Rockingham, *Impact of New Energy Sources, 339* and the general discussion in such studies as that by the Harvard Business School (R. Strobaugh and G. Yergen *Energy Future, 398*). See, more generally, RG, G. In the United States the International Co-generation Society Inc. (800 17th St, Washington DC 2006) acts as a clearinghouse for information.

25 If electricity is produced from fossil fuels there is a net fuel saving; if it is produced from nuclear energy or even photo-voltaic cells, there is utilization of otherwise waste heat. See, for example, G.N. Hatsopoulos, et al., *Capital Investment to Save Energy, 155*, and RG, G.2.

26 Leighton and Kidd Limited, *Report, 220*.

27 A study is now being conducted in Hearst, Ontario, on the feasibility of production of electrical energy from wood wastes and on the institutional arrangements that would be required; see S.N.C. consultants, *Hearst Wood Waste Energy Study, 357*.

28 G.N. Hatsopoulos, et al., *Capital Investment to Save Energy, 155* and RG, G.2.

29 Middleton Associates, *Alternatives to Hydro's Generation Programme, 254*.

30 For example, the study uses the impact of peak shaving, energy conservation, and load shifting in the industrial sector concluded as feasible by Ontario Hydro in 1974 to arrive at a figure of an 8 per cent reduction in peak industrial demand through more efficient use of electricity in the industrial sector. This figure is from three to six times smaller than the one we have quoted, but it is this latter range that is indicated by a wide array of post-1974 studies

of a more and less conservative stripe. Again, the study only assumes an additional 70 megawatts from small hydro sources, whereas there is reason to believe that from ten to thirty times this amount is potentially available and will become increasingly economically attractive over the next decade. Similarly, the study assumes that as much as 750 megawatts could be made available from small generating stations utilizing biomass fuel, but in its interim report *A Race against Time*, *354*, the Royal Commission on Electric Power Planning calls for a demonstration wood-fired generating station of approximately twice that capacity (see p. 180). Assumptions about the rates of commercial penetration of solar technology and co-generation technology are equally conservative by comparison with realistic rates under suitable government legislative and tax incentives.

31 Quoted from D.B. Brooks, *Zero Energy Growth*, *28*. See also W. Leiss, *Quality of Life*, *222*; L. Nader and S. Beckerman, *Quality and Style of Life*, *262*; M. Diesendorf, *Energy and People*, *83*. For social studies in the United States on the impact of the 1973 oil crisis see, for example, M.E. Olsen and J.A. Goodnight, *Social Aspects of Energy Conservation* (Pacific Northwest Regional Commission, Vancouver, Wa., 1977) and R. Perlman and R.L. Warren, *Families in the Energy Crisis* (Cambridge, Mass.: Ballinger, 1977). For the conservation philosophy applied critically to agriculture, see P. Ehrensaft, 'Canadian Agriculture and the Political Economy of the Biosphere,' *Alternatives* 5 (1976), 36–52; J.-R. Mercier, *Énergie et agriculture: le Choix écologique* (Paris: Édition Debard, 1978); W. Barrie, *The Unsettling of America* (San Francisco: Sierra Club, 1977).

CHAPTER 8

1 For Canada, see V.R. Puttagunta, *Temperature Distribution*, *328* and R.K. Swartman, *Alternative Power Generation Technologies*, *403*; Ontario Hydro, *Energy Utilization and Electricity*, *291*; Ontario Ministry of Energy, *Energy Review*, *307* (but see notes 2 of Chapter 3 and 21 of Chapter 7 reporting on conflicts among the sources). For the United States, see, for example, M.H. Ross and R.H. Williams, *Potential for Fuel Conservation*, *347*; and the treatment in A.B. Lovins, *Soft Energy Paths*, *231*.

2 See Ontario Ministry of Energy, *Ontario's Energy Future*, 302, p. 10.

3 The absence of such obvious conserving practices in North America that are widespread in other countries underlines the profligate use of energy in the past. On the use of off-peak load to smooth demand curves see J. Platts, *Electrical Load Management*, *325*, and B.M. Mitchell, et al., *Peak-Load Pricing*, *256*. Seasonal storage facilities with electrical backup can smooth

annual demand curves, though seasonal backup makes variable renewable resources (radiant solar, wind) even more attractive. For discussions of the impact of different energy technologies on employment kinds and levels see RG, L.5 and note 1 of Chapter 5, 10 and 20 of Chapter 7, and note 4 below.

4 Ironically, in the short term, effective insulation plus effective passive solar pickup may so reduce heat demand (to one-quarter or one-fifth of former needs) as to make active solar collection in conjunction with passive techniques unattractive. However, in the medium term of twenty or so years with more readily available, cheaper, and more reliable storage systems, especially seasonal storage systems, combined with decreasing unit prices for active solar units, it can be anticipated that in most parts of Canada active solar collection will become the preferred source for residual heat demand. To the extent it is not cost competitive to retrofit existing buildings to these standards, active solar and other backup fuels will have a larger role to play. For exploratory building designs see the discussions in *Canadian Renewable Energy News* and R. Argue, *The Sun Builders*, *4*; N. Nicholson, *Harvest the Sun*, *274*; *Solar Energy Building*, *275*; R.G. Stein, *Architecture and Energy*, *395*; and R. Argue, *The Well-Tempered Home: Energy Efficient Building for Cold Climates*.

5 See Ontario Hydro, *Energy Utilization and Electricity*, *291*. The important point is that in an appropriately designed, adequately insulated house there is less need for air conditioning than in poorly constructed buildings. The 'climate control' practised on a vast scale in today's buildings is a consequence of poor design and cheap construction, transferring energy costs to future occupiers and ultimately to society at large.

6 H. Swain, et al., *Canadian Renewable Energy Prospects*, *402*, p. 7.

7 P. Middleton, et al., *Renewable Energy Resources*, *253*.

8 K.G.T. Hollands and J.F. Orgill, *Potential for Solar Heating*, *172*.

9 For details of the PUSH program and other related programs see, for example, J.E. Gander and F.W. Belaire, *Energy Futures*, *123*, and the review by A.M. Schrecker in *Conserver Society Notes*, 1 (fall 1978), pp. 35–6 and *Financial Post*, July 15, 1978. Solar technology development, especially in Canada and the United States, is regularly reviewed in *Canadian Renewable Energy News*. Researchers at Sydney University in Australia have developed a new highly efficient solar collector capable of producing temperatures of 200°C. It is expected to be marketable at costs lower than those for 1979 flat-plate collector prices when it becomes commercially available in about five years; see Sydney University *Gazette*, 3 (May 1979), 6. These output temperatures make the device suitable for meeting both low- *and* medium-grade heat demands, thereby reversing the long-standing view that direct solar collection could not meet any but low-grade heat requirements. This is but one of many examples

reinforcing our view that, within broad limits, which technology in a given field becomes economically attractive depends upon the sequence of investments in research, development, and deployment.

10 D. Chapman, *Taxation and Solar Energy*, *46*.

11 See especially the report by Hooper and Angus Associates, *Development of the Solar Heating Alternatives*, *180*; M.K. Berkowitz, *Implementing Solar Energy*, *19*; P.N.L. Battelle, *Federal Incentives Used to Stimulate Energy Production*, *10*; and A.R. Wallenstein, *Barriers and Incentives to Solar Development*, *444*.

12 H. Swain, et al., *Canadian Renewable Energy Prospects*, *402*, p. 7.

13 IBI Group, *Solar Heating*, *184*, p. 12.

14 See, for example, Ontario Ministry of Transportation and Communications, *Utilization of Methanol-based Fuels*, *308*; and, more generally, RG I.4; see also the discussion by D. Elliot and A. Palm-Leiss, *Alternative Fuels*, *98*.

15 See, for example, J. Goldenberg, *Brazil*, *130*; and J. Goldenberg, et al., *Alcohol from Plant Products*, *131*; see Wallace, *Brazilian National Alcohol Programme*, *443*. In the case of Brazil there is a real possibility that expansion of the biomass fuel system may be at the expense of food production, but this remains to be determined. In any case, Canada does not face the same situation since in fact most of the land that would be suitable for biomass production is not suitable for food production.

16 Fisheries and Environment Canada, *Liquid Fuels from Renewable Resources* (Ottawa, May 1978). See also Energy, Mines and Resources Canada, *Tree Power*, *103*.

17 Ontario Ministry of Energy, *Liquid Fuels*, *303*.

18 We also note that the Ontario study evidently simply assumes that the necessary synthetic oils or natural gas feedstocks will be available. Perhaps; but the risks inherent in the policy ought to be made explicit.

19 Increases in the efficiency of the methanol production processes are achievable through use of solar energy to provide process heat and hydro-electric power to provide process hydrogen, from the electrolysis of water, which might increase the yield of methanol from 38 to 50 per cent wood-to-alcohol conversion. In addition, experiments are now under way in the United States to use the stillage from this process as an animal feed and in turn convert the resulting animal manure to methane, thus recovering a further portion of the energy. See M. Wayman, *Opportunities for Fuel*, *446*, for other suggestions for improvements in production efficiency.

20 D. Mackie and R. Sutherland, *Methanol in Ontario*, *237*.

21 TEIGA, *Design for Development*, *309*; and *Long Term Outlook*, *311*.

22 See Ontario Hydro, *Generation – Technical*, *292*, Section 2.2.1.1.

23 See references in note 28 of Chapter 2.

24 TEIGA, *Design for Development*, *309*; and *Long Term Outlook for Ontario*, *311*.

25 See note 10 of Chapter 9.

26 Electric generation owned/operated by local commissions

Region	Municipality	1974 average generation
Eastern	Almonte	668
	Bancroft	213
	Campbellford	1,971
	Eganville	263
	Ottawa	7,867
	Renfrew	1,417
Georgian Bay	Bracebridge	1,688
	Orillia	12,977
	Parry Sound	1,052
Total		27,916

Source: Ontario Hydro, *Energy Utilization and Electricity*, *291*, Table
6.1–3.

Privately owned industrial utility systems

	Capacity	Source
Gananoque Electric Light and Water Co. Ltd.	2,900 customers in Gananoque and vicinity 11 MW	5 hydraulic, 4 thermal
Great Lakes Power Corporation	Sault Ste Marie and vicinity (including industry) 215 MW	3 hydraulic
Canadian Niagara Power	Fort Erie, Nabisco Co., and Canadian Carborundum	
Cornwall Street Railway, Light & Power Company and the St. Lawrence Power Company	15,000 customers in Cornwall and vicinity 55 MW	Power from Hydro Quebec
Abitibi Paper Co. Ltd.	Iroquois Falls and Smooth Rock Falls	Electricity from Abitibi Mill by 125 kV cable
Ontario-Minnesota Pulp & Paper Company Limited	Fort Francis 8 MW	Electricity supplied by company
Huronian Co. Ltd. (wholly owned by INCO)	Sudbury and Algoma 67 MW (3,000 customers plus INCO)	5 hydraulic (49 MW) 29 MW back pressure steam turbines utilizing steam from waste heat recovery

Source: Ontario Hydro, *Energy Utilization and Electricity*, *291*, Section 6.1.4.

27 Personal communication to R.M. from Dr R. Wiley, *Small Scale Hydro Project*, New York State Energy Research and Development Authority (N.Y.S.E.R.D., 2, Rockefeller Plaza, Albany, New York 12223).
28 Royal Commission on Electric Power Planning, *A Race against Time, 354*, p. 180.
29 Bay of Fundy, *Re-assessment of Fundy Tidal Power, 11*; R.H. Clark, *Power from the Tides, 51*; W. Clark, *Energy for Survival, 52*; D. Ross, *Energy from the Waves, 344*; and B. Sorensen, *Renewable Energy, 387*.
30 This estimate by Templin was given to the Ottawa Solar Energy Conference in 1975 and are discussed by A.B. Lovins *Exploring Energy-Efficient Futures, 229*, from which the quote in the text following is taken (p. 14). R.J. Templin, *Wind Energy, 411* and wind studies by C.K. Brown and D.F. Womb, *Analysis of Potential for Wind Energy, 33*, and C.K. Brown, *Preliminary Assessment for Wind Generators, 32*.
31 For other general studies on wind energy see RG, 1.5.
32 See *Canadian Renewable Energy News*, June/July and August 1979.
33 The reduction in international oil shipping associated with the recent economic slowdown, resulting in the mothballing of hundreds of tankers, is illustrative of the reduction in oil spill potential that would result from a permanent conservation-induced reduction in international demand for oil. Similarly, the economic slowdown in Ontario has led to the deferral of thousands of megawatts of nuclear- and fossil-fuel generating capacity; again, illustrative of the permanent removal of this intrusive industrial infrastructure that would result from a conservation-induced reduction in electrical demand.
34 The combustion of carbon rich biomass leads to the production of carbon dioxide. However, if biomass is managed on a renewable basis, the removal of carbon dioxide from the air by the growth of new biomass should more than balance the carbon dioxide added through the burning of methanol.
35 TEIGA, *Design for Development, 309*; and *Long Term Outlook for Ontario, 311*.
36 See, for example, J. Laxer, *The Energy Poker Game, 215* in connection with the urban/hinterland relation between the U.S. and Canadian economies.
37 J. Benson, *Prepared Statement, 17*, and Environmentalists for Employment, *Jobs in Energy, 110*. See also B. Hannon, *Prepared Statement, 147*, and the review by B. McGregor, *Energy, Jobs and the Economy, 247*.
38 About fifteen thousand civil servants are employed in the nuclear industry; see Royal Commission on Electric Power Planning, *A Race against Time, 354*, p. 129.

CHAPTER 9

1 The limitations to market and adversarial models of human interaction are perhaps intuitively clearer in the personal domain. For example, marriages

that are restricted to these two models of interpersonal relationships – increased production/consumption or adversarial bargaining – notoriously suffer from shallowness, limitedness, and stress. It is perhaps unusual in our present cultural climate to speak about such matters as the quality of marriage relationships in the context of public policy. But we believe that adequate public policy simply must recognize that the quality of our interpersonal relationships is the stuff of existence and that rich relationships may not follow the patterns offered by the present public decision-making paradigms. The market and adversary relationships have no wider validity in the energy policy setting than they do in marriage, and it cannot be expected that society can arrive at satisfying public institutions based on them. There is a frightening futility to advocating that we try to recoup in some special 'private' sphere all the interpersonal integrity and richness that must everywhere be forgone in our public lives.

The sense of unease concerning the divorce between the public and the private in economic matters is now widespread in the literature, running from the conservative T. Scitovsky, *The Joyless Economy, 376*, to the more radical H. Henderson, *Creating Alternative Futures, 164*. See also RG, K and L.3. We may follow the difficulties from economics into the theory of decision-making where we find them expressed as formal cautions about constraints of present theories (see the Preface to C.A. Hooker, et al. *Decision Theory, 174*), as intuitive explications of alternative conceptions that seem more adequate to the nature of persons (e.g. G. Vickers *Weakness of Western Culture, 439*; *Value Systems, 440*) or as virulent attacks on the moral and social consequences of the present approaches (e.g. as in L. Tribe *Technology Assessment, 424*). From there the problems may be pursued in almost any direction, into political theory, (see, for example, C.B. MacPherson, *Political Theory of Possessive Individualism, 239* and *Democratic Theory, 240*; and G. Vickers, *Freedom in a Rocking Boat, 438*), into the theory of science (for example, B. Easlea, *Liberation and the Aims of Science, 91*; C.A. Hooker, *Explanation and Culture, 179*), into technology (for example, J. Ellul, *Technological Society, 99*), social science (for example, E. Becker, *Denial of Death, 13*, and *Structure of Evil, 14*) and organization theory (for example, R. Thayer, *End to Hierarchy, 413*). The issues are focused in, and connect the works of, for example, C.A. Hooker, *Human Context, 178*, J. Robertson, *Sane Alternative, 337*, and R. Unger, *Knowledge and Politics, 426*. In our view they point to the relevance of literatures in the field of personal and family development, see W.R. Coulson, *Sense of Community, 67*, R. Greenleaf *Servant Leadership, 136*, C. Rogers *Becoming a Person, 341*, G. Stewart and C. Starrs, *Gone to-day, Here Tomorrow, 396*.

2 See Ontario Ministry of Energy, *Ontario's Energy Future, 302*, for example at p. 40. Note that under the label 'Renewable' the document includes fusion

energy and thorium/plutonium fission cycle as well as geothermal and solar-based energy technology – cf. note 1 to Chapter 6.

3 See note 9, Chapter 1 and cf. note 27, Chapter 4.

4 The conflict over short-term profits and long-term supplies in natural gas discussed in note 8 of Chapter 1 is a case in point. A recent court case in Halifax, Nova Scotia, has revealed the manoeuvering of Exxon during the 1973 oil embargo to increase its own profits at the expense of Canadians (see *Globe and Mail*, Tuesday, October 2, 1979). In this connection, the U.S. Department of Energy has so far charged 35 of the largest oil refiners with $6.4 billion of overcharges (see *London Free Press*, November 7, 1979). Canadians should also recall Exxon's refusal to supply additional Venezuelan crude oil during recent Middle East cutbacks (see *Globe and Mail*, March 14, 1979). These examples are but a suggestion of the gap between private and public interests in the energy field. This is not to deny that there is also a large common interest – the mutual desire for survival. See further discussion on pp. 197–9.

5 The federally created crown corporation Petro-Canada is widely defended, even by Conservative governments in those provinces of Canada vulnerable to manipulation by oil companies, as a 'window on the oil corporations.'

6 See our discussion on p. 199.

7 Hydro's assumptions about its own planning and 'consumer sovereignty' are revealed in its own planning processes; see Ontario Hydro, *Long-Range Planning of the Electrical Power System*, Report 556SP, (February 1974) and *Submission to the Ontario Energy Board – Systems Expansion Program*, 4 vols. (December 1973). In the past, Hydro has evidently tended to make the corresponding assumption that the demand for electric energy is more or less independent of utility policy. Hydro's mandate is specified in *The Power Corporation Act*, (Toronto: Queen's Printer, 1975).

8 See Task Force Hydro, *Hydro in Ontario: A Future Role and Place* (Toronto: Queen's Printer, 1972). An account of Hydro's attempt to remain free of government intervention can be found in H.V. Nelles, *Politics of Development*, 272.

9 We emphasize *official* contact. There have been, and are, many informal, unofficial contacts between Hydro personnel and those of various public agencies and no doubt this has substantially affected the relationship between Hydro and government activities generally. The Royal Commission on Electric Power Planning reports favourably on the co-operation received from Ontario Hydro personnel. But there seems to be little official contact. Yet it is official contact that is the point at which public accountability can be brought to bear, for there is no way to assess the unofficial, informal dimensions to institu-

tional behaviour. There is no systematic way even to decide whether the ad hoc and partial nature of such informal relationships ultimately serves more to hinder or to enhance the public effectiveness of such institutions as Ontario Hydro. We may add that since 1975 Ontario Hydro has taken a greater interest in developing local participatory groups in those areas its impending expansion activities will affect; however, these moves too are at Hydro initiative and within Hydro constraints. See our notes 16 and 27 below.

10 The advantages of large-scale plants are argued on economic grounds. Not only are these grounds uncertain, but it remains to be shown that present cost structures reflect the true social costs of energy production decisions. Moreover, an erstwhile argument to the effect that small plants with uncertain output could not be connected to a provincial energy grid without lowering reliability and increasing eventual capital costs has in effect now been reversed, and many small plants with fluctuating outputs may in fact be now interconnected to increase the overall reliability of the system; see A.P. Rockingham, *Impact of New Energy Sources*, 339. These considerations apply equally to wind generators, electrical plants, and to local co-generation systems.

11 See Ontario Hydro, *Long-Range Planning of the Electric Power System*, Report 556SP (February 1974). Electric solar cells are indeed still expensive and poorly developed, which is not surprising considering the negligible amounts invested in their research and development. But whereas widespread use of solar electricity may be someway off – though even now it is beginning to be seriously contemplated (see *Canadian Renewable Energy News*, June/July and August, 1979) – solar heating certainly is not; in fact, solar houses are appearing all over Canada and in other countries.

12 The main emphasis of Hydro has not so much been on cheaply providing the small amounts of electrical power required by households but rather on satisfying the large industrial customers who account collectively for roughly 40 per cent of the provincial electricity consumption. Through the low cost of electrical power resulting from the strongly regressive rate structure, industrial energy use has become exceedingly wasteful. A rather similar situation prevails in the commercial sector; the wastefulness of the lighting, heating, and cooling systems of large office buildings is notorious.

   Hydro's (largely token) propaganda for energy conservation seems to be directed mainly at the average Ontario household. But although the intellects and wills of the populace have surely to be engaged on this problem, it is not electric toothbrushes and colour TVs that call for crash programs in nuclear plant construction. (Home appliances are estimated to account for 1 per cent of Canadian energy demand.) Opportunities for energy conservation in the industrial and commercial sectors are far more significant than those centred around the household, as is substitution of sources of low-grade heat for high-grade electricity.

13 Ontario Hydro, *Socio-economic Factors*, *294*.
14 *Ibid.*, p. 4.1–5.
15 Ontario Hydro, *Public Participation*, *293*, p. 1-25.
16 See Royal Commission on Electric Power Planning, *A Race against Time*, *354*, pp. 65, 73, 96–7, 101, 155. M. Ouchterlony reviewed Ontario Hydro's recent efforts at public participation in her report to the Royal Commission on Electric Power Planning, *A Review of Ontario Hydro's Public Participation Programme* (Toronto, February 1979). With respect to the federal nuclear authority see comments by Nichols (as briefly discussed on p. 181) and those in *Half Life* (Goderich, Ontario, 1977), by the Ontario Coalition for Nuclear Responsibility. On the values of nuclear electric decision-makers generally and public participation, see D.J. Crossley, *Energy Policy in Australia*, *69*; M. Crow, *The Energy Crisis*, *70*; R.J. Kalter and W.A. Vogeley, *Energy Supply and Government Policy*, *200*; C. Ryan, *Choices*, *358*, A.J. Surrey, et al., *Democratic Decision Making*, *401*; and D. White, et al., *Seeds for Change*, *452*. On the structure and functioning of federal nuclear agencies see the excellent discussion in Chapter 12 of the Interim Report of the Royal Commission on Electric Power Planning, *A Race against Time*, *354*.
17 Ontario Ministry of Energy, *Submission*, *301*, p. 1–1.
18 Ontario Ministry of Energy, *Ontario's Energy Future*, *302*, pp. 67–8, 73, 79, 125–6.
19 TEIGA, *Submission*, *310*, p. 3.
20 TEIGA, *Submission*, *310*, p. 3.
21 See also note 27 below.
22 Ontario Energy Board, *Bulk Power Rates for 1976*, *289*.
23 We note that there is also a federal environmental impact assessment office and a formal assessment process the office oversees which may in future play a substantial role in such national technologies as nuclear energy. To date, however, its role has been decidedly limited; see e.g. R. Lang, *Environmental Impact Assessment*, *212*. Social impact assessment and technology impact assessment processes are being developed that hold substantial promise as new forms for expressing collective evaluation and policy-making, but they are presently without an appropriate institutional framework. For details see C.A. Hooker, *Human Context*, *178*, F. Tester, *Proceedings*, *412* and the extensive references therein.
24 Ontario Legislature, *A New Public Policy Direction for Ontario Hydro*, *299*, p. II-12.

25 K.G. Nichols, *Public Participation in Energy Policy in Canada*, Organisation of Economic Co-operation and Development (Paris, 1978). The report, though it has not been formally released by OECD, was described in the *Globe and Mail*, June 29, 1978, and reviewed in *Conserver Society Notes*, 1, 3 (1978), p. 33.

26 David Brooks, former director of the Federal Office of Energy Conservation and author of several important studies cited herein, is now director of Ottawa Energy Probe and the Canadian representative of Friends of the Earth International. He is using his offices to act as co-ordinator of alternative energy groups across Canada of which there are now more than twenty provincial conservation/renewable energy groups affiliated under this umbrella organization. For details see *Alternatives* 8 (summer/fall 1979).

The point made in the text is illustrated afresh in the situation for solar energy technology. On the one hand, there is tremendous fragmentation of development, with many individual companies competing with one another, more or less wholly out of touch with each other and, more important, out of touch with the other institutions most likely to make their products relevant (e.g., architects, developers, planners). Most government officials and members of the public are unaware of the activities of the members of the Solar Energy Society of Canada Inc. Until a private citizen with specialized experience in a private organization (Energy Probe) produced a book (*The Sun Builders*, by Robert Argue), there had been no systematic attempt to co-ordinate the experiences and insights of those many dozens of people across Canada who have more or less independently built solar houses on an experimental basis. (Though there had been other, individual private initiatives, e.g. that by Nicholson reflected in *Harvest the Sun*, 274.) Until a private group, the Bureau of Municipal Research, made a preliminary study of energy use and conservation at the municipal level, little or no attempt had been made to involve municipal and regional officers in energy policy discussions. (We acknowledge here both the support of the Royal Commission for Electric Power Planning for the Middleton Associates study in Oakville and local hearings of the Commission throughout Ontario as important, if limited, additions to these efforts.) The result is an ad hoc, fitful, and frustrating situation where prediction is hazardous, management non-existent, and the constraints of ignorance, disorganization, and lack of support frustrating to all. We hold the view that the difficulties we have identified in institutional design and function stem from the way in which such market-oriented institutions reflect the assumptions and imperatives of a market-dominated political economy and culture, but we do not press the point here. Such views are elaborated in a report to the Ontario Royal Commission of Electric Power Planning, *Institu-*

tions, 173, and applied in *Meaning of Environmental Problems, 176.* Instead, we turn to a discussion of alternatives.

27 It is easy for people to lose sight of how very large decisions with a far-reaching social impact can be made by only a handful of people in highly centralized institutions. We might begin with columnist Hugh Winsor's remark that the Chairman of Ontario Hydro, appointed in December 1978, was 'a close personal friend and advisor' of the Ontario premier, and notes that it was the Chairman's brother 'when he was Minister of Energy, Economics and Development, who bequeathed his former position as Vice-Chairman of Hydro' to the present Ontario premier who was then a back-bencher; *Globe and Mail,* December 18, 1978. Little more than a month later Winsor pointed out that during the Ontario Energy Board Hearings In The Matter of Principles of Power Costing and Rate Making Appropriate For Use By Ontario Hydro, the Deputy Minister of Energy had set up a private lunch between the chairman of the Ontario Energy Board, the chairman of the Association of Major Power Consumers (AMPCO), and the vice-chairman of Ontario Hydro. Winsor remarked: 'About a week after the lunch, Ontario Hydro dropped from its proposal before the Energy Board two elements that had been concerning AMPCO, including something called the Large User Pricing Rule which would have created a surcharge on new plants. [Later] the crown corporation submitted something called a "line management study" which turns the whole nine-volume study submitted originally by Hydro upside-down. It has the effect of maintaining the status quo, including the cross-subsidy to the big users.' The Ontario Hydro vice-president 'did admit on cross-examination that what amounts to a fundamental shift in policy was not discussed with or ordered by the Board of Directors of Hydro.' *Globe and Mail,* January 24 and January 25, 1979.

This approach to public policy-making can infect even the most well-intentioned public exercise. For example, the Royal Commission on Electric Power Planning held an all-day meeting devoted to the topic of public participation in energy policy-making and implementation on January 8, 1978. In the middle of that day, ironically, the chairman of the Commission went to lunch with, among others, the chairman of Ontario Hydro, the deputy pro-secretary of the Natural Resources Secretariat, and the presidents of the Universities of Toronto and Western Ontario, who have held important positions in relation to energy policy and research development; everyone else at the meeting concerned with public participation was excluded from the lunch.

There is also the opposite effect of the refusal to transfer information among interested institutions, often motivated by political conflict. There was the time, for example, when officials of AECL simply refused to testify in person before the Select Committee of the Ontario Legislature investigating Ontario Hydro, though they did agree to answer questions in writing; see *Globe and Mail*, July 12, 1978. Subsequently, the chairman of AECL again refused to appear, insisting that it was not 'in the public interest'; see *Globe and Mail*, August 22, 1978. And there was the time when Ontario Hydro and the Atomic Energy Control Board accused each other of thwarting the implementation of further safety systems on Ontario nuclear power plants while, it seems, in the preceding year neither body had moved significantly to assist the implementation; *London Free Press*, October 3, 1979. The same day's newspaper revealed rivalry of opinion between Ontario Hydro and the Ontario Ministry of Energy, the former insisting that the province is committed to at least a study of a post-Darlington nuclear power station, with the latter denying it. Considering the potential importance of the commitment – as against competing uses for the resources to develop alternative energy systems – this is a matter of some public importance.

Even without these 'curtains of privacy' it is extremely difficult for the general public to obtain an accurate picture for decision-making. A good example of the difficulties is given by the matter of reactor safety. Mr Pease, an expert at AECL, indicated that a figure for the likelihood of a major accident could be derived from a U.S. study, yielding a chance between 1 in 4,000 and 1 in 100,000 per reactor per year, while Mr Morrisson of Ontario Hydro insisted that a more reasonable figure was 1 in 1,000,000 per reactor per year. The difference is rather important, but impossible for anyone save a few experts to check. (Under the worst case, the 1 in 4,000 figure, it suggests a major accident every 40 years on average for 100 Canadian reactors), Testimony before the Royal Commission on Electric Power Planning has been carefully discussed by R. Torrie and G. Edwards, *Summary Argument*, *423*. As M. Diesendorf, secretary of the Australian Society for Social Responsibility in Science, has pointed out: 'Already in 1964 we suffered the incineration in the upper atmosphere of a space vehicle containing 17,000 curies (1 kg) of plutonium-238 as a heat source. Before launching, "expert" scientists had tried to convince Dr K.Z. Morgan, former president of the International Radiation Protection Association, that the chances of such an accident occurring were one in 10 million' (*Sydney Morning Herald*, 15 September 1977). We remind the reader that nuclear power is not unique in this respect.

28 For those who find it helpful, we could put the point this way: learning involves tolerance of failure, but the failures must be confined to minimize their costs; that is, learning requires 'safe-fail' (safe-for-failure) or resilient

systems, not 'fail-safe' (safe-from-failure) systems. But centralized massive technologies and bureaucracies tend overwhelmingly to be designed as fail-safe systems. They make decisions or deliver technologies to their entire constituencies from some central source, so failure at the source tends to be catastrophic and is usually wholly unacceptable. Ontario Hydro and its nuclear-electric system are excellent examples of a fail-safe technology requiring a matching fail-safe decision-making process. By contrast, decentralized institutions contain failure locally, permitting learning compatibly with overall social continuity.

Once again nuclear energy technology, or energy technology in general, is not the only area subject to centralization and fail-safe rigidity. Central Canada is steadily centralizing its population, industrial production, energy production, and so on. In these circumstances, there is rapidly increasing contact between the human population and potentially hazardous sections of the energy-industrial structure. This inevitably calls for increasingly rigid fail-safe designs, because of the increased costs of failure, and leads us steadily to increasing police powers and decreasing public participation – all in the name of the public interest. After the rail accident in Mississauga, Ontario, police were congratulated for the effectiveness with which they handled the forced evacuation of large numbers of people, and there were calls for more stringent security measures and wider police emergency powers. Few people evidently stopped to notice how many hard-won civil liberties can so easily be forgone in this manner.

29 In Theobold's terms (*TEGS 1994, 416*), it leads to the emergence of a sapiential authority, the authority willingly accorded to the intelligently competent person by others who recognize and trust that competence, to replace authority based on power and exclusionary expertise; in Greenleaf's terms (*Servant Leadership, 136*), it offers a widening scope for servant leadership.

30 For the emerging new communications systems based on the micro-electronics 'revolution' of the past decade, see GAMMA, *Report on the Information Society*, a report prepared for the Department of Communications (Montreal: Université de Montréal, March 1979); C. Gotlieb and A. Borodin, *Social Issues in Computing* (New York: Academic Press, 1973); the papers by Mowshowitz and Valaskakis/Martin in C.A. Hooker, *Human Context, 178*; A. Mowshowitz, *The Conquest of Will: Information Processing in Human Affairs* (Reading, Mass.: Addison-Wesley, 1976); A. Mowshowitz (ed.), *Proceedings of the Second IFIP Conference on Human Choice and Computers* (Amsterdam: North-Holland, 1979).

But one should be very cautious about the form these technologies may take, for not only may they be used to create vast new inequities of power,

they may also suffer from severe inherent limitations; see in this respect the
excellent discussion of television in J. Mander, *Four Arguments for the Elimination of Television* (New York: Morrow Quill 1979).

31 Human wars hinge on divorcing the opposing people, as people, from oneself
by using intervening abstractions such as 'enemy' or 'gook' and by divorcing
oneself from one's deeds by using intervening abstractions such as 'pre-
emptive elimination' and 'pacification.' Adequate conflict resolution must
bring the self and its morality back into touch with both reality and with a
reality-serving use of language.

32 See the account of the Eneraction programme in *Canadian Renewable Energy
News* (December 1978/January 1979). More recently, the Ontario Ministry of
Transportation and Communications, co-operating with the Ontario Ministry
of Energy, sponsored its own meeting for municipal and regional officers on
energy conservation opportunities at the local level but, as a commentary on
the difference between grass roots and bureaucratically driven exercises, the
public were not invited to the meeting and the press were actively excluded;
see *London Free Press*, November 24, 1979.

33 See note 30 above.

34 There are already some experiments in regional government in Ontario, and
the provincial government seems to have accepted this line of development,
at least for the long run. As we understand it, the present experiments suffer
from several difficulties: regional governments have been superimposed upon,
but not integrated with, the municipal governments; they have been undertaken in the absence of the full measure of institutional support envisaged
here; and they share the financial inadequacies of the municipal level. Nonetheless, they constitute experiments that may be worthwhile building upon.

35 The centre would work through the day-to-day contacts available in the centre
as well as with the direct in-house means discussed earlier. Consumer and citizen interests are represented in the energy offices by membership in the governing councils and perhaps in other component groups as well. Industrial/
commercial interests are similarly represented in the offices, but in view of the
social importance and technical intricacy in this area a special liaison group
seems called for; hence the Industrial Energy Council. It would be the responsibility of this council to translate the proposed energy policy into its industrial/
commercial implications and vice versa. This information, fed to the remainder
of the office and then out to local government and citizens' groups would be an
important but by no means the only component in the iterative process of
forming energy policy.

Since neighbourhood centres would permit many other professional groups
to operate in the same setting – medical, dental, and other health care person-

nel, various branches supporting welfare (Family and Children's Services, etc.) – the opportunity exists to discuss the implications of energy policy in each of these areas in an immediate and local setting with professionals in the areas concerned. Correspondingly, it offers professional people a chance to deal directly with the complex, but real issues facing communities rather than being confined, as they are on many occasions, by their institutionalized professional roles.

36 Alternatively these experiments might only demonstrate a tendency for the energy offices to expand into local centralized energy monopolies. Their production and distribution interests might come to dominate their policy formation responsibilities, perhaps even subverting the network of critical feedback from local government and citizens. Recalling our attitude to such matters from earlier discussion, we would then recommend divesting the offices of production and distribution responsibilities.

37 There would, however, need to be more guarantees given of publicly criticizable procedures and decisions than the current Ontario Environment Impact legislation offers. We note that a provincial energy impact assessment group would, at the least, have to work closely with the developing environmental impact branch of the provincial Ministry of the Environment, and it should likewise maintain constructive relations with those involved in the closely related social impact assessment and technology assessment fields – indeed, co-ordination of these efforts might well be its major function. See also note 23 above.

38 We expect direct secretariat-secretariat communication, paralleling inter-regional communication (e.g., between the energy secretariat and that of treasury), direct group-group communication (e.g., as illustrated, but also between inter-ministerial groups), and indirect ministry-ministry communication via various cabinet committees or secretariats. Communication with the public would take place predominantly through the local energy offices.

39 Finally, we stress that it would be a mistake to separate the existing civil service attached to ministerial portfolios from this decision-making process. Though we speak of a distinction between the ministerial civil service and criterial institutions, this is to emphasize the existence and potential roles of the latter, not to deny the former a place. On the contrary, we have explicitly suggested that the proposed provincial-level energy policy institution should be based on the existing Ministry of Energy and relate appropriately to the relevant cabinet committees and other ministries. Further, there are some developments in the ministerial civil service that provide an important model for criterial institutional functioning, specifically the federal and provincial environmental impact assessment process. Though relatively new and still

developing, and restricted to a specific ministerial context (Ministry of the Environment), the assessment too is an open hearing process, much more explicitly oriented towards future planning and able to focus on evaluative decisions. The impact assessment procedure might well provide the appropriate place to relate the roles of criterial institutions to those of the ministerial civil service in the new policy-making setting we propose.

This might, however, prove too severe a decentralization to cope with the dual need to formulate coherent provincial energy policy, especially with respect to fiscal and national energy policies, and to avoid too great regional inequities, unreasonable local biases, and local domination by national or international enterprises. Achieving a good balance will be a very complex matter and must surely be arrived at through considerable experimentation and public evaluation. Just how the local-provincial representation is best managed must be matters for urgent further study, experimentation, and debate.

40 Some aspects of this issue have been discussed in C.A. Hooker and R. van Hulst, *Institutions*, *173*, Chapter 6.

41 Haunting and frustrating examples of this have been documented in W. Stewart, *Paper Juggernaut* (Toronto: McClelland & Stewart, 1979), and D. Halberstam, *The Best and the Brightest* (New York: Fawcett, 1972). It is well known that parliament has come to depend increasingly on what we referred to as the direct ministerial civil service. The days are long gone when parliamentarians made most of the detailed public decisions affecting even their own constituencies, let alone those of the country. Most of these decisions are now delegated to the civil service in one or other of their regulatory roles. Equally, the civil service develops policy statements; by the time these emerge at cabinet level they are like the tips of icebergs, reflecting in their very structure and concepts a vast underlying framework of theories, calculations, assumptions, and conceptualizations that are largely opaque to those engaged in the policy discussions. Finally, governments are increasingly bombarded by special-interest lobbyists who bring to their activities both a single-purpose focus that no parliamentarian can afford to have and specialized and often 'inside' information that is often beyond the reach of anyone in government.

In these circumstances it is not surprising that the role of parliament has been becoming less rationally manageable and less meaningful. The emergence in the last half century of both a vast civil service and a plethora of commissions, boards, crown corporations, and the like may be seen as an attempt to contain these problems within some democratic framework by delegating them to public or quasi-public institutions under the nominal control of parliament. There is a variety of reasons why virtually all of this machinery has taken a centralized form (e.g., simplicity of design, need to deal with international business, the inherent outlook of the market paradigm). The point is that, in its own way, this machinery too has often tended to undercut parliamentary roles, directly through its opaqueness to parliamentary inquiry and indirectly through its effective exclusion of relevant public participation.

42 Royal Commission on Electric Power Planning, *A Race against Time*, *354*, p. 129.

43 Not all of us are agreed on the relative emphasis we would place on these three poles or on the scope to be given to pluralism. Moreover, just as parliament may be located among complementary institutional alternatives, so may the market operate in a context of exchanges, bartering, direct communal production, and distribution. Whether these latter activities are to be classified as 'private enterprise' or as 'local participation' is ultimately unimportant, just as it is not important precisely what are our own individual preferences. What is of importance is that the public have clearly before them the options actually open to them, are able to learn about the consequences of pursuing them, and are able to express their choices among them both individually and socially.

44 See, for example, N.J. Lucas, *Local Energy Centres*, *235*; J. Robertson, *Sane Alternatives*, *337*; V. and R. Routley, *Social Theories*, *352*; R. Thayer, *End to Hierarchy*, *413*; R. Theobald, *Habit and Habitat*, *415*; E.F. Schumacher, *Small Is Beautiful*, *367*, E.F. Schumacher and P.N. Gillingham, *Good Work*, *368*; G. Stewart and C. Starrs, *Gone Today, Here Tomorrow*, *396*.

CHAPTER 10

1 On health policy as a commodity, see C.A. Hooker and R. van Hulst, *Meaning of Environmental Problems*, *176*.

2 We have not had anything explicit to say in this book about a general theory of policy formation and content. We have been content to illustrate what is required in the specific case of energy. An approach to a new theory of policy is under development by C.A.H., and some aspects of its formulation appear in a Ph.D. thesis by Dr A. Hanna, 'A Theoretical Framework for the Evaluation of Public Policy' (University of Western Ontario, 1980).

3 For the interested reader we offer the following references as useful places to begin. S. Beer, *Platform for Change* (New York: Wiley, 1975); R.J. Bennett and R.J. Chorley, *Environmental Systems* (London: Methuen 1978); C.W. Churchman, *Challenge to Reason* (New York: McGraw-Hill, 1968); E. Jansch, *Design for Evolution* (New York: Braziller, 1975); E. Jansch and C.H. Waddington (eds.), *Evolution and Consciousness* (London: Addison-Wesley, 1976);

R. May, *Stability and Complexity in Model Ecosystems* (Princeton, NJ: Princeton University Press, 1973); H.B. Thorelli, *Strategy and Structure Performance* (Bloomington, Indiana: Indiana University Press, 1977).

4 This circumstance is unique to the twentieth century and of crucial institutional importance. The situation and its consequences are discussed in C.A. Hooker, *Human Context, 178*, and *Explanation and Culture, 179*.

Reader's guide to the bibliography

What follows is a reader's guide to the bibliography assembled under a few of the many topics of interest. No attempt has been made to evaluate articles. The sections are anything but mutually exclusive, and remarks at the ends of sections direct readers' attention to other closely related sections. Italicized listings represent our suggestions for where an interested novice might begin.

Note: As this book went to press the Ontario Royal Commission on Electric Power Planning announced its intention to publish its final report in 9 volumes; these, too, should be an important source of information in all areas.

A. GOVERNMENT DOCUMENTS

A.1. Canadian: Federal 19, 28, 30, 41, 49, 81, *100*, 101, 102, 103, 104, 105, 107, 108, 119, 120, *123*, 140, 191, 192, 238, 246, 253, 321, 377, 389, 402, 419.

A.2. Canadian: Ontario 84, 98, 237, 289, 297, 298, 299, 300, 301, *302*, 303, 304, 305, 306, 307, 308, 309, 310, 311, 345, 383, 396.

A.3. United States 17, 45, 46, 56, 98, 147, 263, 269, 270, 281, 329, 346, 391, 430, 431, 433, 434.

A.4. Royal Commissions 331, 354, 356.

A.5. Other Public Canadian Sources 5, 6, 11, 26, 27, 64, 154, 188, 213, 290, 291, 292, 293, 294, 295, 296, 328, 355, 370, 371, 372, 373, 374, 417, 425, 455, 460.

A.6. Royal Commission on Electric Power Planning in Ontario 9, 63, 78, 167, 169, 173, 175, 180, 184, 220, 222, 254, 255, 260, 288, 339, 354, 355, 403, 423 – see also 9 vol. final report of the Commission (Toronto, 1980).

B. WORLD ENERGY 104, 112, 114, 160, 205, 317, 322, 340, 357, 359, 445, 458, *459*, 461 (Energy Systematics: 282, 375).

C. PRIVATE ENERGY CORPORATIONS 23, 24, 106, 217, 334, 360, 382, 405.

D. CANADIAN FOSSIL FUELS 25, 216, 238, 327, 353, 419, 445.

E. CANADIAN ELECTRICITY 256, 290, 291, 295, 298, 299, 301, 325, *354*, 391.

F. NUCLEAR ENERGY
F.1. General 5, 6, 18, 21, 38, 40, 43, 44, 73, 96, 113, 121, 138, 156, 170, 181, 198, 204, 214, 244, 245, 259, 261, 264, 279, 280, 287, 288, 313, 314, 331, 335, 336, *354*, 356, 409, 420, 436, 442, 448, 449, 450. See also L.
F.2. Radiation 35, *94*, 109, 111, 171, 185, 193, 194, 195, 241, 243, 258, 270, 326, 351, 369, 378, 407, 429, 430, 433, 437, 460.
F.3. Waste Disposal 95, 102, 425.
F.4. Risk (including cost/benefit assessment) 15, 60, 116, 122, 125, 128, 137, 187, 188, *230*, 266, 281, 286, 388, 390, 392, 421, 422, 423, 427, 428, 447. See also F.6, I.8, L.1, L.6.
F.5. Economics 9, 36, 89, 207, 208, 223, *231*, 234, 252, 296, 316, 371.
F.6. Institutional 85, 93, 148, *173*, 202, 248, 250, 251, 271, 293, 363. See also B, F.4, M.

G. CONSERVATION
G.1. General 7, 27, 28, *29*, 56, 64, 76, 79, 133, 139, 144, 147, 157, 203, 219, 228 229, 231, 254, 260, 319, 321, 324, 347, 364, 365, 366, 372, 384, 386, *398*, 406, 414, 431, 435, 452, 453, 454. See also J.2, H.1, L.3.
G.2. Co-generation 139, 166, 220, 453, 456.

H. ENERGY DEMAND
H.1. Structure and Projections 30, 31, 47, 48, 54, 84, 117, 118, 134, 141, 168, 219, 238, 242, 285, 291, 328, 338, 345, 391, 417, 419, 434. See also G.1, L.1.
H.2. Urban Form 49, 72, 81, 107, 162, 169, 209, 225, 342, 393, 395.
H.3. Community Energy Audits 64, *105*, 190, 213, 255, 277, *284*, 319.

I. RENEWABLE ENERGY
I.1. General 52, *62*, 189, 227, 231, 232, 253, 387, 417. See also J.2, N, L.5.
I.2. Electrical Interconnection 199, 339.
I.3. Solar *19*, 46, 75, 88, 119, 120, 158, 161, 172, 180, 184, 196, 211, 304, 305, 332, 334, 385, 402, 403, 432, 444, 452, 455, 457. Practical guides: 2, 4, 274, 275.
I.4. Biomass 3, 8, 42, 98, 103, 130, 131, 182, 191, 192, 237, 308, 383, 443, 446.
I.5. Wind 32, 33, 400, 411, 455.
I.6. Waves/Tidal 11, 51, 344.

I.7. Hydro-electric See A, E.
I.8. Environmental Impact 12, 34, 108, 210, 315, 343, 389. See also F.4.

J. APPROPRIATE TECHNOLOGY
J.1. General 82, 152, 246, 267, *367*. See also L.6.
J.2. Soft Energy Paths 62, 63, 231, 232, 268, 276, 380, 381, 397, 410. See also G.1, I.1.

K. ENERGY AND ALTERNATIVE ECONOMICS 55, 61, 74, *126*, 127, *164*, 165, 312, 318, 367, 415, 451. See also N, L.3.

L. ENERGY POLICY
L.1. General 60, *61*, 66, 69, 70, 100, 112, 115, 117, 118, 123, 135, 142, 160, *163*, 200, 215, 216, 224, 226, 227, 228, 231, 249, 295, 301, 302, 330, 338, 348, 353, *354*, 358, 359, 370, 372, 374, *398*, 404, 461. See also A.
L.2. Socio-political 7, 60, 113, 124, 143, 151, 154, 167, *175*, 197, 210, *221*, 222, 236, 263, 271, 272, 288, 294, 312, 458. See also L.6.
L.3. Economic Policy 10, 27, 45, 46, 65, 74, 78, 79, 87, 101, 126, 129, 145, 150, 154, 168, *173*, 206, 249, 256, 335, 376, 441.
L.4. Centralization 78, 92, 250, 265, 272, 278.
L.5. Employment 17, 60, 110, 146, 147, *247*.
L.6. Ethics and Values 16, 55, 58, 74, 83, 125, 132, 145, 151, *177*, *186*, 227, 230, 248, 261, 262, 263, 273, 283, 324, 334, 358, 363, 408, 424, 448, 449, 450. See also L.2, J.1, L.1, N, K.
L.7. Environmental 57, *86*, 134, 153, 201, 212, 221, 236, 267, 323, 412.

M. ENERGY INSTITUTIONS/DECENTRALIZATION 39, 68, 69, 173, *176*, 183, 235, 320, 329, 394, 399, 401, 452. See also N, L.4, F.6.

N. COMMUNALIST PHILOSOPHY AND THEORY 13, 14, 53, 67, 91, 99, 136, 140, 149, 159, 178, 179, 186, 239, 240, 267, 337, 341, 349, 350, 352, 361, 368, 379, 396, 413, 416, 418, 424, 426, 438, 439, 440. See also L.1, K, J.1, L.6.

Bibliography

1 Alexander, T., 'Industry can save energy without stunting its growth,'
   Fortune (May 1977).
2 Anderson, B., *The Solar Home Book: Heating, Cooling and Designing with the
   Sun*. Harrisville, NH: Chesshire Books, 1976.
3 American Petroleum Institute, *Alcohols: A Technical Assessment of their Application as Fuels*. Washington, DC: American Petroleum Institute, July 1976.
4 Argue, R., *The Sun Builders*, Toronto: Renewable Energy in Canada, 1978.
5 Atomic Energy of Canada Limited, *Nuclear Power: The Canadian Issues*.
   Chalk River: Atomic Energy of Canada Ltd., 1977 (AECL-5800).
6 – *Final Argument Relating to the Canadian Nuclear Power Program*. Chalk
   River: Atomic Energy of Canada Ltd, 1978 (AECL-6200).
7 Bailey, C., 'Environmentalism, the Left, and the Conserver Society.'
   Conserver Society Notes, 1 (summer 1978) 22–7.
8 Ban, W.J. and F.A. Parker, *Introduction of Methanol as a New Fuel into the
   U.S. Economy*. A Report for the American Energy Research Company. San
   Diego, Calif.: Foundation for Ocean Research, March 1976.
9 Banerjee, S. and L. Waverman, *Life Cycle Costs of Coal and Nuclear Generat-
   ing Stations*. A report prepared for the Royal Commission on Electric Power
   Planning. Toronto, July 1978.
10 Battelle, Pacific Northwest Laboratories, *An Analysis of Federal Incentives
   Used to Stimulate Energy Production*. Richland, W.A.; 1978.
11 Bay of Fundy, Power Review Board, *Re-assessment of Fundy Tidal Power*.
   Ottawa, Fredericton and Halifax, November 1977.
12 Beall, S.E., et al., *An Assessment of the Environmental Impact of Alternative
   Energy Sources*. Oakridge, Tenn.: Oakridge National Laboratory, September
   1974 (ORNL-5024).
13 Becker, E., *The Denial of Death*. New York: The Free Press, 1968.

14 – *The Structure of Evil*. New York: The Free Press, 1976.
15 Beer, S., 'Questions of Metric,' in S. Beer, *Platform for Change*. New York: Wiley, 1975.
16 Benson, J., *Energy and Reality: Three Perceptions*. Fairfax, Va.: Institute for Ecological Policies, 1977.
17 – *Prepared Statement*. A statement prepared for Hearings before the Sub-committee on Energy of the Joint Economic Committee, Congress of the United States, March 1978. Washington, DC: U.S. Government Printing Office, 1978.
18 Berger, L.M., *Nuclear Power: The Unviable Option*. Palo Alto, Calif.: Ramparts Press, 1976.
19 Berkowitz, M.K., *Implementing Solar Energy Technology in Canada: The Costs, Benefits, and the Role of Government*. A Report for the Renewable Energy Resources Branch, Energy, Mines and Resources, Canada. Ottawa: Queen's Printer, 1977 (VI77-7).
20 Berry, W., *The Unsettling of America: Culture and Agriculture*. San Francisco: Sierra Club, 1977.
21 Bethe, H.A., 'The Necessity of Fission Power,' *Scientific American*, 234 (1976), 21.
22 Birre, A., *Une Politique de la terre*. Marcq-Ville (France): Vie et Action, 1967.
23 Blair, J.M., *The Control of Oil*. New York: Vintage, 1978.
24 Bregha, F., *Bob Blair's Pipeline*. Toronto: J. Lorimer, 1979.
25 Bresee, P. and S. Tyler, 'Alberta's Athabasca Oil Sands: A Canadian Perspective,' *Alternatives*, 4 (1975), 21–32.
26 Britton, N.H. and J.M. Gilmour, *The Weakest Link: A Technological Perspective on Canadian Industrial Underdevelopment*. Ottawa: Science Council of Canada, October 1978.
27 Brooks, D.B., *Economic Impact of Low Energy Growth in Canada: An Initial Analysis*. Ottawa: Economic Council of Canada, Discussion Paper #126, December 1978.
28 – 'Zero Energy Growth for Canada: Necessity and Opportunity,' Energy, Mines and Resources, Canada, Office of Energy Conservation, Paper 8, May 1976.
29 – 'A Real Option: Conservation to 1990 and beyond,' *Alternatives*, 7, 1 (fall 1977).
30 Brooks, D.B., R. Erdman, and G. Winstanley, 'Some Scenarios of Energy Demand in Canada in the Year 2025,' A Study for the Energy Review Group, Energy, Mines and Resources Canada, April 20, 1977.
31 Brooks, H. and J.M. Hollander, 'United States Energy Alternatives to 2010 and Beyond: The CONAES Study,' *Annual Reviews of Energy*, 4 (1979), 1–70.

32 Brown, C.K., *Preliminary Assessment of the Potential for Large Wind Generators as Fuel Savers in A.C. Community Diesel Power Systems in Ontario*. Toronto: Ontario Research Foundation, 1976.

33 Brown, C.K. and D.F. Womb, *An Analysis of the Potential for Wind Energy Production in Northwestern Ontario*. Toronto: Ontario Research Foundation, 1975.

34 Budnitz, R.J. and J.P. Holdren, 'Social and Environmental Costs of Energy Systems,' *Annual Reviews of Energy*, 1 (1976), 553–80.

35 Bujdoso, E. (ed.), *Proceedings of the International Radiological Protection Association, Second European Congress on Radiation Protection: Health Physics Problems of Internal Contamination*. Budapest: Acadèmiai Kiadò, 1973.

36 Bupp, I.C., J.C. Derian, et al., 'The Economics of Nuclear Power.' *Technology Review*, 77 (February 1975), 15–38.

37 Bureau of Municipal Research, *What Can Municipalities Do about Energy?* Toronto, March 1978.

38 Caldicott, H., *Nuclear Madness*. Melbourne, Australia: Jacaranda Press, 1978.

39 Caldwell, L.K., 'Energy and the Structure of Social Institutions.' *Human Ecology*, 4 (1976), 31–45.

40 Canadian Coalition for Nuclear Responsibility, 'Nuclear Power: The Deadliest Gamble,' *Alternatives*, 7 (fall 1977), 26–33.

41 Canadian Environmental Advisory Council, *Reports of the First and Second Meetings of Public Interest Groups with the Canadian Environmental Advisory Council*. Ottawa: Queen's Printer, 1978.

42 Canadian Morbark Ltd., *Bio-Sun Energy and Ontario's Forests*. North Bay, Ont.: Canadian Morbark Ltd., 1978.

43 Canadian Nuclear Association, *Nuclear Power in Canada: Questions and Answers*. Toronto: Canadian Nuclear Association, 1975.

44 – *The Role of Nuclear Power in Ontario*. Submission to the Royal Commission On Electric Power Planning. Toronto, 1976.

45 Chapman, D., *Prepared Statement*. A statement prepared for hearings before the Sub-committee on Energy of the Joint Economic Committee, Congress of the United States, March 1978. Washington, DC: U.S. Government Printing Office, 1978.

46 – *Taxation and Solar Energy*. A report for the Solar Energy Office, California Energy Commission, June 1979.

47 Chapman, P. *Fuels Paradise*. Harmondsworth: Penguin, 1975.

48 Chesshire, J.H. and A.J. Surrey, *Estimating U.K. Energy Demand for the Year 2000: A Sectoral Approach*. Sussex: Science Policy Research Unit, University of Sussex, February 1978.

49 Chibuk, J., *Urban Form and Energy: A Selected Review*. A report prepared for the Ministry of State for Urban Affairs. Ottawa, July 1977.

50 Cina, C. and T. Goldfarb, 'Three Mile Island and Nuclear Power,' *Science for the People*, 11 (July/August 1979), 10–18.
51 Clark, R.H., 'Power from the Tides,' *GEOS* (fall 1978), 12–14.
52 Clark, W., *Energy for Survival*. New York: Anchor (Doubleday), 1975.
53 Cleveland, H. and T.W. Wilson, Jr., *Humangrowth: An Essay on Growth, Values and the Quality of Life*. Princeton, NJ: Aspen Institute Program in International Affairs, 1978.
54 Cole, H.S.D., C. Freeman, M. Jahoda, and K.L.R. Pavitt, *Thinking about the Future: A Critique of the Limits to Growth*. London: Chatto & Windus, 1973.
55 Collard, D., *Economics and Altruism: A Study in Non-selfish Economics*. New York: Oxford University Press, 1978.
56 Committee on Nuclear and Alternative Energy Systems (CONAES), 'U.S. Energy Demand: Some Low Energy Futures,' *Science*, 200 (April 1978) 142–52.
57 Commoner, B. *The Closing Circle*. New York: A.A. Knopf, 1971.
58 – *Energy and Human Welfare: A Critical Analysis*, 3 vols. New York: Macmillan Information, 1975.
59 – 'Energy and Labour: Job Implications of Energy Development or Shortage,' *Alternatives*, 7 (summer 1978), 4–13.
60 – *The Politics of Energy*. Boston: Knopf, 1979.
61 – *The Poverty of Power: Energy and the Economic Crisis*. New York: Bantam, 1977.
62 Conway, C., 'Soft Energy Paths,' *Alternatives* 8 (summer/fall 1979) and 9 (spring 1980). These issues are devoted to a report outlining Soft Energy Strategies for the ten provinces of Canada.
63 Conway, C., R. Crow and P. Szegedy-Marezak, *Energy Planning in a Conserver Society: The Future's Not What It Used to Be*, Submission to the Royal Commission on Electric Power Planning, February 1978.
64 Conway, C., *Conservation: Options for Canadian Electrical Utilities*. A background paper for the Cities Energy Conference, Toronto, January/February 1980. (Available from Office of Conservation, Toronto Hydro.)
65 Cook, P.L., 'Energy Policy Formulation,' in W. Leontief (ed.), *Structure, System and Economic Policy*, Cambridge: Cambridge University Press, 1977.
66 Cook, P.L. and A.J. Surrey, *Energy Policy: Strategies for Uncertainty*. London: Martin Robertson, 1977.
67 Coulson, W.R., *A Sense of Community*. Columbus, Ohio: Chas. Merrill, 1973.
68 Craig, P., J. Darmstadter and S. Rattien, 'Social and Institutional Factors in Energy Conversion,' *Annual Reviews of Energy*, 1 (1976), 535–51.
69 Crossley, D.J., *Energy Policy in Australia: The Social/Institutional Context and Procedures for Policy Formulation*. Brisbane, Australia: Griffith University, School of Australian Environmental Studies, working paper 6/78.

70 Crow, J., 'The Energy Crisis: A Solution-Multiplying Approach,' *Social Alternatives*, 1 (September 1979) 54–6.

71 Cubberley, D., 'Science on Trial,' *Last Post*, 7 (November 1979), 14–20.

72 Curtis, F. (ed.), *Considerations and Opportunities for Energy Conservation in Urban and Regional Planning: Conference Proceedings*. Queen's University, Kingston, Ont., March 1979.

73 Curtis, R. and E. Hogan, *Perils of the Peaceful Atom*. New York: Ballantyne, 1969.

74 Daly, H.E., *Steady-State Economics: The Economics of Bio-physical Equilibrium and Moral Growth*. San Francisco: W.H. Freeman, 1977.

75 Daniels, F., *Direct Use of the Sun's Energy*. New York: Ballantyne, 1964.

76 Darmstadter, J., J. Dunkerley and J. Alterman, *How Industrial Societies Use Energy: A Comparative Analysis*. Baltimore, MD: Johns Hopkins University Press, 1977.

77 – 'International Variations in Energy Use: Findings from a Comparative Study,' *Annual Reviews of Energy*, 3 (1978), 201–24.

78 Dean, J.O., *The Role of Ontario Hydro as an Economic Development Tool of the Province*. Report prepared for the Ontario Royal Commission on Electric Power Planning. Toronto, March 1976.

79 Deane, W. (ed.), *Growth in a Conserving Society*. Toronto: Yorkminster Publishing, 1979.

80 Dennison Mines Limited, *The 'Uranium Cartel Myth.'* Toronto: October 19, 1977.

81 Desfor, G. and M. Miloff, *Urban Form and Energy Consumption* (draft Technical Report No. 1: Confirmation of Urban Form Factors), a report prepared for the Ministry of State for Urban Affairs on behalf of Middleton Associates. Toronto: Middleton Associates, September 1978.

82 Dickson, D., *Alternative Technology and the Politics of Technological Change*. Glasgow: Fontana/Collins, 1974.

83 Diesendorf, M. (ed.), *Energy and People: Social Implications of Different Energy Futures*. Canberra, Australia: Society for Social Responsibility in Science, 1979. (SSRS Box 48, O'Connor, A.C.T., Australia 2601.)

84 Dillon, R.M., 'Ontario: Toward a Nuclear Electric Society?' An address to the Canadian National Energy Forum, October, 1975.

85 Doern, G.B., *The Atomic Energy Control Board*. Ottawa: Law Reform Commission of Canada, 1977.

86 Dorf, R.D. and Y. Hunter, (ed.), *Appropriate Vision: Technology, the Environment and the Individual*. San Francisco: Boyd & Fraser, 1978.

87 Downs, J.R. *The Availability of Capital to Fund the Development of Canadian Energy Supplies*. Calgary: Canadian Energy Research Institute, 1977.

88 Duffie, J.A. and W.A. Beckman, *Solar Energy Thermal Processes*. New York: Wiley Inter-Science, 1974.

89 Duncan, S., 'Energy? Canada's Economy? Think Big. Ottawa Is,' *Financial Post*, March 25, 1978.

90 Earthscan, *Whose Solar Power?* London: July 1979. (10 Percy Street, London W1P ODR, England).

91 Easlea, B., *Liberation and the Aims of Science*. London: Chatto & Windus, 1973.

92 Easterbrook, W.T. and H.G. Aitken, *Canadian Economic History*. Toronto: Macmillan, 1956.

93 Ebbin, S. and R. Kasper, *Citizen Groups and the Nuclear Power Controversy: Uses of Scientific and Technological Information*. Cambridge, Mass.: MIT Press, 1974.

94 Edwards, G., *Estimating Lung Cancers*. Halifax: Canadian Coalition for Nuclear Responsibility, March 1978.

95 – *Nuclear Wastes: What, Me Worry?* Montreal: Canadian Coalition for Nuclear Responsibility, 1978.

96 Eggleston, W., *Canada's Nuclear Story*. Toronto: Clarke, Irwin, 1965.

97 Ehrensaft, P., 'Canadian Agriculture and the Political Economy of the Biosphere,' *Alternatives*, 5 (no. 3/4, 1976), 36–54.

98 Elliot, D. and A. Palm-Leis, *Alternate Fuels for Vehicular Use*. Toronto, Ministry of Transportation and Communications, Research and Development Division, 1977.

99 Ellul, J., *The Technological Society*. New York: Vintage Books, 1967.

100 Energy, Mines and Resources Canada, *An Energy Strategy for Canada: Policies for Self-reliance*. Ottawa: Queen's Printer, 1976.

101 – *Financing Energy Self-reliance*. Ottawa: Queen's Printer, 1977 (EP 77-8).

102 – *The Management of Canada's Nuclear Wastes*. A report by L.K. Hare et al., Ottawa: Queen's Printer, August 1977 (EP77-6).

103 – *Tree Power: An Assessment of the Energy Potential of Forest Bio-mass in Canada*. A Report by P. Love and R. Overend for the Renewable Energy Resources Branch, Energy, Mines and Resources, Canada. Ottawa: Queen's Printer, 1978 (ER78-1).

104 – *World Uranium Requirements in Perspective*. A Report by R.M. Williams, Assistant Advisor, Uranium Energy Policy Sector. Ottawa: Queen's Printer, July 1978 (ER78-4).

105 – *Municipal Energy Audit* Ottawa: Conservation and Renewable Energy Branch, Energy, Mines and Resources Canada, January 1980 (draft).

106 Engler, R., *The Brotherhood of Oil: Energy Policy and the Public Interest*. Chicago: University of Chicago Press, 1977.

107 Environment Canada, *Energy Conservation through Land-Use Decisions*. A report prepared by the IBI Group, Toronto. Ottawa: Environment Canada, March 1976.

108 – *James Bay Hydro-electric Project: A Statement of Environmental Concerns and Recommendations for Protection and Enhancement Measures*. Ottawa: Environment Canada, March 1975.

109 Environmental Policy Institute, *Proceedings of a Congressional Seminar on Low-Level Ionizing Radiation*. Washington, DC, July 1977.

110 Environmentalists for Full Employment, *Jobs and Energy*. Washington, DC, 1977.

111 Epstein, S., *The Politics of Cancer*. San Francisco: Sierra Club, 1978.

112 Erickson, B.W. and L. Waverman, *The Energy Question: An International Failure of Policy*. 2 vols. Toronto: University of Toronto Press, 1974.

113 Flood, M. and R. Grove-White, *Nuclear Prospects: A Comment on the Individual, the State, and Nuclear Power*. London: Friends of the Earth, undated.

114 Flower, A.R., 'World Oil Production.' *Scientific American*, 238 (March 1978), 42–9.

115 Folley, G., *The Energy Question*. Harmondsworth: Penguin, 1976.

116 Ford, D.F. and H.W. Kendall, *An Assessment of the Emergency Core Cooling Systems Rule Making Hearings*. San Francisco: Friends of the Earth (for Union of Concerned Scientists), 1974.

117 Ford Foundation, Energy Policy Project, *Exploring Energy Choices, A Preliminary Report*. Washington DC: Ford Foundation, 1974.

118 – *A Time to Choose*. Final Report, Cambridge, Mass.: Ballinger, 1974.

119 Foster, H.D. and W.D. Sewell, *Canadian Perspective on United States Solar Policies: An Examination of U.S. Government Solar Policies to Accelerate the Growth of Solar Manufacturing*. Renewable Energy Resources Branch, Energy, Mines and Resources, Canada. Ottawa: Queen's Printer, 1978 (ER78-3).

120 – *Solar Home Heating in Canada: Problems and Prospects*. Report 16, Office of the Science Advisor, Fisheries and Environment Canada. Ottawa: Fisheries and Environment Canada, 1977.

121 Francis, J. and P. Abrecht, (eds.) *Facing up to Nuclear Power*. Edinborough: St. Andrew Press, 1976.

122 Fuller, J.G., *We Almost Lost Detroit*. New York: T.Y. Crowel Co. (for Reader's Digest Press), 1975.

123 Gander, J.E. and F.W. Belaire, *Energy Futures for Canadians: Long-Term Energy Assessment Program* (LEAP). Report of a study prepared for Energy, Mines and Resources Canada. Ottawa: Queen's Printer, 1978.

124 Gardner, J. and J. Marsh, 'Recreation in Consumer and Conserver Societies,' *Alternatives*, 7 (summer 1978), 25–9.

125 Geesaman, D.P. and D.E. Abrahamson, 'The Dilemma of Fission Power,' *Bulletin of the Atomic Scientist*, 30 (November 1974), 37–41.

126 Georgescu-Roegen, N., 'Energy and Economic Myths.' *Southern Economic Journal*, 41 (1975), 347–81.

127 – *The Entropy Law and the Economic Process.* Cambridge, Mass.: Harvard University Press, 1971.

128 Gillinsky, V., 'Plutonium, Proliferation and Policy,' *Technology Review*, 79 (no. 4, February 1977), 58–66.

129 Goldberg, M.A., 'The Economics of Limiting Energy Use,' *Alternatives*, 1 (1972), 3–13.

130 Goldemberg, J., 'Brazil: Energy Options and Current Outlook,' *Science*, 200 (April 1978), 158–64.

131 Goldemberg, J. and J.R. Moreira, *Alcohol from Plant Products: A Brazilian Alternative to the Energy Shortage.* Pre-print IFUSP/P-126, September 1977 (available from Universedade de Sãopaulo, Instituto de Fisica).

132 Goldsmith, E., 'The Future of an Affluent Society: The Case of Canada.' *The Ecologist*, 7 (1977), 161–94.

133 Goldstein, D. and A. Rosenfeld *Projecting an Energy-Efficient California.* Berkeley: Energy and Environment Division, University of California, 1975.

134 Goldstein, W., 'The Political Metaphors of Environmental Control.' *Alternatives*, 2 (1973), 11–17.

135 Greenberger, M. and R. Richels, 'Assessing Energy Policy Models: Current State and Future Directions,' *Annual Review of Energy*, 4 (1979), 467–500.

136 Greenleaf, R., *Servant Leadership: A Journey into the Nature of Legitimate Power and Greatness.* New York: Paulist Press, 1977.

137 Greenwood, T., et al., *Nuclear Proliferation: Motivations, Capabilities and Strategies for Control.* Washington, DC: Council on Foreign Relations, 1977.

138 Groupe des Scientifiques pour l'Information sur l'Énergie nucléaire, *Electro-nucléaire: Danger.* Paris: Seuil, 1976.

139 Gyftopoulos, E.P., L.J. Lazaridis, and T.F. Widmer, *Potential Fuel Effectiveness in Industry.* Cambridge, Mass.: Ballinger, 1974.

140 Haas L., *Renewable Energy for Developing Countries: A Preliminary Assessment of the Potential and Canadian Capability,* A report prepared for the Conservation and Renewable Energy Branch, Energy, Mines and Resources, Canada and the Canadian International Development Agency by Middleton Associates. Toronto, March 1979.

141 Häfele, W. and W. Sassin, *Science, Technology and Natural Resources: Resources and Endowments; An Outline of Future Energy Systems.* Laxenburg,

Austria: International Institute for Applied Systems Analysis, 1978. A paper prepared for the NATO Science Committee's 20th Anniversary Commemoration Conference, Brussels, 1978.

142 Hambraeus, G. and S. Stillesjo, 'Perspectives on Energy in Sweden,' Palo Alto, Calif.: *Annual Review of Energy*, 2 (1977), 417–54.

143 Hammarlund, J.R. and L.N. Lindberg, *The Political Economy of Energy Policy: A Projection for Capitalist Societies*. Madison: Institute for Environmental Studies, University of Wisconsin-Madison, December 1976.

144 Hannon, B., 'Energy Conservation and the Consumer,' *Science*, 180 (July 1975), 95–102.

145 – *Energy, Growth and Altruism*. Urbana-Champaigne: Center for Advanced Computation, University of Illinois, 1975.

146 – 'Energy, Labor, and the Conserver Society.' *Technology Review*, 79 (March/April 1977), 47–53.

147 – *Prepared Statement*. A statement prepared for hearings before the Sub-committee on Energy of the Joint Economic Committee, Congress of the United States, March 1978. Washington, DC: U.S. Government Printing Office, 1978.

148 Hardin, G., 'Living with the Faustian Bargain,' *Bulletin of the Atomic Scientists*, 32 (November 1976), 25–9.

149 Harding, J.A., 'Ecology as Ideology,' *Alternatives*, 3 (1974), 18–22.

150 Harding, W., *Nukenomics: The Political Economy of the Nuclear Industry*. Regina: Regina Group for a Non-Nuclear Society, 1979.

151 – *Uranium: Correspondence with the Premier*. Regina: Regina Group for a Non-Nuclear Society, 1979.

152 Harper, P. and G. Boyle, (eds.), *Radical Technology*. London: Wildwood House, 1976.

153 Harte, J. and A. Jassby, 'Energy Technologies and Natural Environments: The Search for Compatibility.' *Annual Reviews of Energy*, 3 (1978), 101–46.

154 Hartle, D., 'Federal-Provincial Relations and the Canadian Energy Issue,' (in) Economic Council of Ontario, *Energy Policies for the 1980's: And Economic Analysis*. Toronto, in press.

155 Hatsopoulos, G.N., E.P. Gyftopoulos, R.W. Sant, and T.F. Widmer, 'Capital Investment to Save Energy,' *Harvard Business Review*, 56 (1978), 111–12.

156 Hayes, D., *Nuclear Power: The Fifth Horseman*. Washington, DC: Worldwatch Institute, 1976. (Worldwatch Institute: 1776 Massachusetts Avenue, NW, Washington, DC 20036.)

157 – *Energy: The Case for Conservation*. Washington, DC: Worldwatch Institute, 1976.

158 – *Energy: The Solar Prospect*. Washington, DC: Worldwatch Institute, March 1977.

159 – *Energy for Development: Third World Option*. Washington, DC: Worldwatch Institute, December 1977.
160 – *Rays of Hope: The Transition to a Post-Petroleum World*. New York: W.W. Norton, 1977.
161 – *The Solar Energy Timetable*. Washington, DC: Worldwatch Institute, April 1978.
162 Heintz, M., *Land Use and Energy Bibliography*. Washington, DC: Energy Research and Conservation Development Corporation, February 1976 (draft).
163 Helliwell, J., 'Canadian Energy Policy,' *Annual Reviews of Energy*, 4 (1979), 175–230.
164 Henderson, H., *Creating Alternative Futures: The End of Economics*. New York: Berkley Wyndhover, 1978.
165 – 'The Transition to Renewable Resource Societies and the End of "Flat Earth Economics,"' *Alternatives*, 8 (1978), 15–22.
166 Herman, S.W. and J.S. Cannon, *Energy Futures: Industry and the New Technologies*. Cambridge, Mass.: Ballinger, 1977.
167 Hill P.G., *The Social Costs of Electric Power Generation*, a report prepared for the Royal Commission on Electric Power Planning, 1977.
168 Hitch, C.J. (ed.), *Modeling Energy-Economy Interactions: Five Approaches*. Washington, DC: Resources for the Future Inc., 1977.
169 Hix, J., *Energy Demands for Future and Existing Land-Use Patterns*. A report prepared for the Royal Commission on Electric Power Planning. Toronto, October 1977.
170 Hodgettes, J.E., *Administering the Atom for Peace*. New York: Atherton, 1964.
171 Holden, C., 'Low-Level Radiation: A High-Level Concern,' *Science*, 204 (April 1979), 155–8.
172 Hollands, K.G.T. and J.F. Orgill, *Potential for Solar Heating in Canada*. A Report Prepared for the Division of Building Research, National Research Council of Canada. Ottawa: National Research Council, 1977.
173 Hooker, C.A. and R. van Hulst, *Institutions, Counter-Institutions and the Conceptual Framework of Energy Policy Making in Ontario*. Toronto: Royal Commission on Electric Power Planning, May 1976.
174 Hooker, C.A., J. Leach and E. McClennen, *Foundations and Applications of Decision Theory*. Dordrecht, Holland: D. Reidel Publishing Co., 1977.
175 Hooker, C.A., 'The Socio-economic Significance of Electric Power Policy,' in [355].
176 Hooker, C.A. and R. van Hulst, 'The Meaning of Environmental Problems for Public Political Institutions,' in [203].
177 Hooker, C.A., 'Value Judgments and Energy Policy,' *Social Alternatives*, 1 (September 1979), 35–8.

178 – *The Human Context for Science and Technology*. A report prepared for the Social Sciences and Humanities Research Council of Canada, 3 vols. London, Ont.: University of Western Ontario, November 1979; edited in 1 vol. and reprinted by the Social Sciences and Humanities Research Council, Ottawa, May 1980.
179 – 'Explanation and Culture,' *Humanities in Science*, Center for the Humanities, University of Southern California, 1980. In press.
180 Hooper and Angus Associates, *Legislative and Governmental Actions Bearing on the Development of the Solar Heating Alternative*. Toronto: Royal Commission on Electric Power Planning, January 1978.
181 Hornby, I. and B. McMullan, *The Nuke Book: The Impact of Nuclear Development*. Ottawa: Pollution Probe, rev. 2nd ed., 1977.
182 Hughes, F.P. and F.L. Hughes, 'The Perpetual Energy Source Canada Ignores Is Waiting to Be Exploited,' *Science Forum*, 60 (December 1977), 8–10.
183 Hunnius, G., 'The Yugoslav System of Decentralisation and Self-management,' in C. Bennello and G. Roussopoulos, (eds.) *The Case for Participatory Democracy* New York: Viking Press, 1971.
184 IBI Group, *Solar Heating: An Estimate of Market Penetration*. A report prepared for the Royal Commission on Electric Power Planning. Toronto, 1977.
185 Ibser, H.W., 'The Nuclear Energy Game: Genetic Roulette,' *The Progressive*, 40 (January 1976), 15–18.
186 Illich, I.V., *Energy and Equity*. London: Calder and Boyars, 1974.
187 Inhaber, H., 'Risk with Energy from Conventional and Non-conventional Sources,' *Science*, 203 (February 1979), 718–23.
188 – *The Risk of Energy Production*. Ottawa: Atomic Energy Control Board, 1978 (AECB 1119).
189 Institute économique et juridique de l'énergie, *Alternatives aux nucléaire*. Grenoble: Presses universitaires de Grenoble, 1978.
190 Institute for Local Self-Reliance, *Planning for Energy Self-reliance: A Case Study of the District of Columbia* Washington, DC, January 1979 (draft).
191 InterGroup Consulting Economists Ltd., *Economic Pre-feasibility Study: Large Scale Methanol Production from Surplus Canadian Forest Bio-mass*, Part I: Summary Report, Part II: Working Papers. Ottawa: Fisheries and Environment Canada, September 1976.
192 – *Liquid Fuels from Renewable Resources: Feasibility Study*. A report prepared for the Government of Canada, Interdepartmental Steering Committee on Canadian Liquid Fuels. Winnipeg: InterGroup Consulting Economists Ltd., May 1978.

193 International Commission on Radiological Protection, *Publication 9*. Oxford: Pergammon Press, 1966.

194 – *Publication 15*. Oxford: Pergammon Press, 1970.

195 Jacobi, W., 'Problems Concerning the Recommendation of a Maximum Permissible Inhalation Intake of Short-Lived Radon Daughters,' in 35.

196 Johansson, T.G. and P. Steen 'Solar Sweden,' *AmBio*, 7 (1978), 70–4.

197 Jones, C.O., 'American Politics and the Organisation of Energy Decision Making,' *Annual Reviews of Energy*, 4 (1979), 99–122.

198 Jungk, R. (E. Mosbacher, transl.), *The Nuclear State*. London: John Colber, 1979.

199 Kahn, E., 'The Compatibility of Wind and Solar Technology with Conventional Energy Systems,' *Annual Review of Energy*, 4 (1979), 313–52.

200 Kalter, R.J. and W.A. Vogely, (eds.), *Energy Supply and Government Policy*. Ithaca: Cornell University Press, 1976.

201 Kelley, D.R., *The Energy Crisis and the Environment: An International Perspective*. New York: Praeger, 1977.

202 Kneese, A.V., 'The Faustian Bargain,' in *Resources*, no. 44. New York: Resources for the Future, September 1973. Reprinted in *112*.

203 Knelman, F.H., *Anti-Nation: Transition to Sustainability*. Oakville, Ont.: Mosaic Press, 1979.

204 – *Nuclear Energy: The Unforgiving Technology*. Edmonton: Hurtig, 1976.

205 – 'The Geo-politics of Oil,' Montreal: Concordia University, undated.

206 Kohr, L., *The Overdeveloped Nations: The Dis-economies of Scale*. New York: Shocken Books, 1978.

207 Komanoff, C., *Comparison of Nuclear and Coal Costs*. New York: Komanoff Energy Associates, 1978.

208 – 'Doing Without Nuclear Power,' *New York Review of Books*, XXVI (May 1979), 14–17.

209 Lackie, J., G. Masters, H. Whitehouse and L. Young, *Other Homes and Garbage: Designs for Self-sufficient Living*. San Francisco: Sierra Club, 1975.

210 Lajambe, H., 'Les Côuts et externalités de l'amenagement hydro-électrique de la baie James.' Thesis, McGill University, Montreal, 1978.

211 – *Helio-Québec: l'autonomie énergetique du Québec dans une perspective écologique*. St-Bruno de Montarville, Quebec, November 1979.

212 Lang, R., 'Environmental Impact Assessment: Reform or Rhetoric?' in *221*.

213 Lang, A. and A. Armour, *Sourcebook: Cities Energy Conference*. Toronto, 1979. (Available from Information Services, Planning and Development Department, City of Toronto, City Hall, Toronto M5H 2N2, together with a *Report* on the conference.)

214 Law, C. and R. Glen, *Critical Choice: Nuclear Power in Canada*. Toronto: Corpus Information Services Ltd., 1978.

215 Laxer J., *The Energy Poker Game: The Politics of the Continental Resources Deal*. Toronto: New Press, 1970.

216 – *Canada's Energy Crisis*, Toronto: James, Lewis & Samuel, 1974.

217 Laxer, J. and A. Martin (eds.), *The Big Tough Expensive Job: Imperial Oil and the Canadian Economy*, Toronto: Press Porcepic, 1976.

218 Leach, G., *Energy and Food Production*. London: International Institute for Environment and Development, 1975.

219 Leach, G., et al., *A Low Energy Strategy for the United Kingdom*. New York: Humanities Press, 1979.

220 Leighton & Kidd Limited, *Report on Industrial Bi-product Power*. Toronto: Royal Commission on Electric Power Planning, May 1977.

221 Leiss, W., *Ecology Versus Politics in Canada*. Toronto: University of Toronto Press, 1979.

222 – *Quality of Life and Energy Conservation*. A report prepared for the Royal Commission on Electric Power Planning. Toronto, October 1977.

223 Leonard and Porteus Ltd., *Economic Impact of Nuclear Energy Industry in Canada: Executive Summary*. A Report for the Canadian Nuclear Association. Toronto: Canadian Nuclear Association, August 1978.

224 Lindberg, L.N., et al. (eds.), *The Energy Syndrome: Comparing National Responses to the Energy Crisis*. Lexington, Mass.: Lexington Books, 1977.

225 Linton K., 'Case History: Energy Conservation,' *The Canadian Architect* (March 1977), 42–7.

226 Lonnroth, M., P. Steen, and T.B. Johansson, *Energy in Transition*. Stockholm: Secretariat for Future Studies, 1977. (Secretariat for Future Studies: P.O. Box S-103, 10 Stockholm, Sweden.)

227 Lovins, A.B. and J.H. Price, *Non-Nuclear Futures: The Case for an Ethical Energy Strategy*. Cambridge, Mass.: Friends of the Earth/Ballinger, 1975.

228 Lovins, A.B., 'Energy Strategy: The Road Not Taken?' *Foreign Affairs*, 1 (October 1976), 65–96.

229 – 'Exploring Energy-Efficient Futures for Canada,' *Conserver Society Notes*, 1 (no. 4, 1976), 5–16.

230 – 'Cost-Risk Benefit Assessments in Energy Policy,' *George Washington Law Review*, 45 (August, 1977), 911–43.

231 – *Soft Energy Paths*. Cambridge, Mass.: Friends of the Earth/Ballinger, 1977.

232 – 'Soft Energy Technologies,' *Annual Reviews of Energy*, 3 (1978), 477–517.

233 – 'A Neo-Capitalist Manifesto: Free Enterprise Can Finance our Energy Future.' *Not Man Apart* (September/October 1978), 8–10.

234 – *Electric Utility Investments: Excelsior or Confetti?* Invited address to the E.F. Hutton Fixed Income Research Conference on Public and Investor Owned Electric Utilities, New York, March 1979. (Available from Friends of the Earth.)

235 Lucas, N.J., *Local Energy Centres.* New York: Applied Science, 1978.

236 Macdonald, R., 'Energy, Ecology, and Politics,' in *221.*

237 MacKay, D. and R. Sutherland, *Methanol in Ontario: A Preliminary Report.* Toronto: Ontario Ministry of Energy, November 1976.

238 MacNabb, G.M. 'The Canadian Energy Situation in 1990'; prepared for the Third Canadian National Energy Forum, Halifax, April 1977.

239 Macpherson, C.B., *The Political Theory of Possessive Individualism.* London: Oxford University Press, 1962.

240 – *Democratic Theory: Essays in Retrieval.* Oxford: Clarendon Press, 1973.

241 Mancuso, T.F., A. Stewart and G. Kneale, 'Radiation Exposures of Handford Workers Dying from Cancer and Other Causes,' *Health Physics,* 33 (1977), 369–85.

242 Marchetti, C., 'Primary Energy Substitution Models: On the Interaction between Energy and Society,' in *Proceedings of the Workshop on Energy Demands, May 1975,* Laxenburg, Austria: International Institute for Applied Systems Analysis, 1976, 803–44.

243 Marx, J.L., 'Low-Level Radiation, Just How Bad Is It?' *Science,* 204 (April 1979), 160–4.

244 Mawson, C.A., L. D'Easum and T. Schrecker, 'An Exchange of Views on Nuclear Power,' *Alternatives,* 4 (1975), 22–7.

245 Mawson, C.A., 'Nuclear Power with Energy Conservation,' *Alternatives,* 2 (summer 1978), 30–4.

246 McCallum, B., *Environmentally Appropriate Technology.* Ottawa: Fisheries and Environment Canada, 1977.

247 McGregor B., 'Energy, Jobs and the Economy,' *Conserver Society Notes,* 1 (no. 1, 1978).

248 McPhee, J., *The Curve of Binding Energy.* Paris: Ferrar, Straus & Girous, 1974.

249 Meadows, D.L. (ed.), *Alternatives to Growth.* I. Cambridge, Mass.: Ballinger, 1977.

250 Mendell, G. and C. Guedeney, *L'Angoisse atomique et les centrales nucléaires.* Paris: Payot, 1973.

251 Metzger, P., *The Atomic Establishment.* New York: Simon and Schuster, 1972.

252 Miller, S., *The Economics of Nuclear and Coal Power.* New York: Praeger, 1976.

253 Middleton, P., R. Argue, T. Burrell and G. Hathaway, *Canada's Renewable Energy Resources: An Assessment of Potential.* Toronto: Middleton Associates, 1976.

254 Middleton Associates, *Alternatives to Ontario Hydro's Generation Program.* A report prepared for the Royal Commission on Electric Power Planning by Peter Victor, Middleton Associates. Toronto, November 1977.

255 – *Energy Management at the Local Level.* A report prepared for the Royal Commission on Electric Power Planning, Toronto, March 1978.

256 Mitchell, B.M., W.G. Manning, Jr, and J.P. Acton, *Peak-Load Pricing: European Lessons for U.S. Energy Policy.* Cambridge, Mass.: Ballinger, 1978.

257 Mitchell, D., *The Politics of Food.* Toronto: J. Lorimer, 1975.

258 Morgan, K.Z., 'Cancer and Low Level Ionizing Radiation,' *Bulletin of the Atomic Scientists,* 34 (September 1978), 30–40.

259 Mueller, P., *On Things Nuclear: The Canadian Debate.* Toronto: Canadian Institute of International Affairs, 1977.

260 Murgatroyd, W., 'Efficient Utilization of Energy,' in [*355*].

261 Myers, D.B., *The Nuclear Power Debate: Moral, Economic, Technical and Political Issues.* New York: Praeger, 1977.

262 Nader, L. and S. Beckerman, 'Energy as It Relates to the Quality and Style of Life,' *Annual Reviews of Energy,* 3 (1978), 1–28.

263 Nader, L. (ed.), *Energy Choices in a Democratic Society.* A report prepared for the Committee on Nuclear and Alternative Energy Systems (CONAES). Washington, DC: National Academy of Sciences, November 1979.

264 Nader, R. and J. Abbotts, *The Menace of Atomic Energy.* New York: W.W. Norton, 1977.

265 Naisbitt, J., *Satellite Power System; A White Paper on Centralization.* Springfield, Va.: National Technical Information Service, 1978.

266 Nash, C., D.W. Pearce and J. Stanley, 'An Evaluation of Cost-Benefit Analysis Criteria,' *Scottish Journal of Political Economy,* 22 (1975), 121–4.

267 Nash, H. (ed.), *Progress as if Survival Mattered.* San Francisco: Friends of the Earth, 1978.

268 – (ed.), *The Energy Controversy: Soft Path Questions and Answers.* San Francisco: Friends of the Earth, July 1979.

269 National Academy of Sciences, U.S., *Report 3.* Committee on Biological Effects of Ionizing Radiation (Chairman: E.P. Radford), Washington, DC, May 1979. (cf. BEIR Report 2, Washington, 1972.)

270 – *Report,* Synthesis Panel, Committee on Nuclear and Alternative Energy Systems (CONAES). Washington, DC, 1978.

271 Nelkin, D. and S. Fallows, 'The Evolution of the Nuclear Debate: The Role of Public Participation,' *Annual Reviews of Energy,* 3 (1978), 275–312.

272 Nelles, H.V., *The Politics of Development: Forest, Mines and Hydro-electric Power in Ontario, 1849–1941*. Toronto: Macmillan, 1974.

273 Newman, B.K. and D. Day, *The American Energy Consumer*. Cambridge, Mass.: Ballinger, 1975.

274 Nicholson, N., *Harvest the Sun*. Ayer's Cliff, Quebec: Renewable Energy Publications, 1978.

275 Nicholson, N. and B. Davidson, *The Nicholson Solar Energy Building Manual*. Ayer's Cliff, Quebec: Renewable Energy Publications, 1977.

276 Norman, C., *Soft Technologies, Hard Choices*. Washington, DC: Worldwatch Institute, June 1978.

277 Northeast Appropriate Energy Technology Network, *Franklin County Energy Study*, Greenfield, Ma., 1978.

278 Novick, S., 'The Electric Power Industry. Part I: Prelude to Monopoly; Part II: Monopoly of Power,' *Environment*, 17 (1975), I: 7–13, II: 32–9.

279 – *The Electric War: The Fight over Nuclear Power*. San Francisco: Sierra Club, 1976.

280 Nuclear Energy Policy Study Group, *Nuclear Power: Issues and Choices*. A Report Sponsored by the Ford Foundation and Administered by the Mitre Corporation. New York: Ballinger, 1977.

281 Nuclear Regulatory Commission, U.S., *Reactor Safety Study*, 9 vols. Study Director: M.C. Rasmussen. Washington, DC: U.S. Government Printing Office, 1975 (WASH-1400).

282 Odum, H.T., *Energy Basis for Man and Nature*. New York: McGraw-Hill, 1975.

283 O'Hearn, J., 'Beyond the Growth Controversy: An Assessment of Responses,' *Alternatives*, 7 (summer 1978), 35–41.

284 Okagaki, A. and J. Benson, *County Energy Plan Guidebook: Creating a Renewable Energy Future*, Fairfax, Va.: Institute for Ecological Policies, September 1979.

285 Olivier, D., *Low Energy Scenario for the United Kingdom, 1975–2025*. A paper prepared for the International Project for Soft Energy Power Conference, 'Transitional Programs towards a Soft Energy Systems on a European Scale,' Rome, May 1979, London: Earth Resources Research, 1979.

286 Olson, M., 'Cost-Benefit Analysis, Statistical Decision Theory, and Environmental Policy,' in F. Suppe and P.D. Asquith (eds.), *PSA 1976*. II. East Lansing, Mich.: Philosophy of Science Association, 1977.

287 Olson, M.C., *Unacceptable Risk: The Nuclear Power Controversy*. New York: Bantam, 1976.

288 Ontario Coalition for Nuclear Responsibility, *Half-Life: Nuclear Power and Future Society*. A report prepared for the Royal Commission on Electric Power Planning. Report prepared by R. Torrie. Goderich, Ont., 1977.

289 Ontario Energy Board, *Ontario Hydro: Bulk Power Rates for 1976*, Part I. Toronto: Queen's Printer, 1976.

290 Ontario Hydro, *Submission to the Select Committee of the Ontario Legislature Investigating Ontario Hydro*, Toronto, 1976.

291 – *Energy Utilization and the Role of Electricity*. Submission to the Royal Commission on Electric Power Planning with respect to the public information hearings. Toronto, April 1976.

292 – *Generation – Technical*. Submission to the Royal Commission on Electric Power Planning with respect to the public information hearings. Toronto, March 1976.

293 – *Public Participation*. Submission to the Royal Commission on Electric Power Planning with respect to the public information hearings. Toronto, March 1976.

294 – *Socio-economic Factors*. Submission to the Royal Commission on Electric Power Planning with respect to the public information hearing. Toronto, 1976.

295 – *Review of the System Expansion and Financial Plans*. Submission to the Royal Commission on Electric Power Planning with respect to the public information hearings. Toronto, April 1977.

296 – *Comparative Costs of Nuclear Generating Plants*. Submission to the Royal Commission on Electric Power Planning, March 1978.

297 Ontario Legislature, *An Act to Establish the Ministry of Energy*, Bill 134, 29th Legislature, Ontario. Toronto: Queen's Printer, 1973.

298 – Select Committee Investigating Ontario Hydro, *A Public Policy Direction for Ontario Hydro: Staff Presentation*. Toronto, April 20, 1976.

299 – Select Committee Investigating Ontario Hydro, *A New Public Policy Direction for Ontario Hydro: Final Report*. Toronto: Queen's Printer, June 1976.

300 – Select Committee on Hydro Affairs Investigating Ontario Hydro's Uranium Contracts with Dennison Mines and Preston Mines. *Summary and Recommendations of the Staff to the Select Committee*. Toronto, February 20, 1978.

301 Ontario Ministry of Energy, *Submission to the Royal Commission on Electric Power Planning*. Toronto, May 1976.

302 – *Ontario's Energy Future*. Toronto: Queen's Printer, April 1977.

303 – *Liquid Fuels in Ontario's Future*. Toronto: Ministry of Energy, Advisory Group on Synthetic Liquid Fuels, May 1978.

304 – *Perspectives on Access to Sunlight*. Toronto, 1978.

305 – *Component Cost of Solar Energy Systems*. A Report by J.T. Orgill and R.M.R. Higgin. Toronto, August 1978.

306 – *Feasibility of Central Heating in Downtown Sarnia*. A report prepared by James F. MacLaren Ltd., London, Ontario. Toronto, March 1979.

307 – *Ontario Energy Review*. Toronto: Queen's Printer, June 1979.

308 Ontario Ministry of Transportation and Communication, *Utilization of Methanol-Based Fuels in Transportation*. A Report for the Advisory Group on Synthetic Liquid Fuels, Ontario Ministry of Energy. Toronto, April 1978.

309 Ontario Ministry of Treasury, Economics and Inter-Governmental Affairs, *Design for Development, Ontario's Future: Trends and Options*. A Statement by the Government of Ontario on Provincial and Regional Development. Prepared by the Regional Planning Branch, Ministry of Treasury, Economics and Inter-Governmental Affairs. Toronto: March 1976.

310 – *Submission to the Royal Commission on Electric Power Planning*. Toronto, May 1976.

311 – *Long Term Outlook for Ontario*; Toronto, June 1976.

312 Ophuls, W., *Ecology and the Politics of Scarcity: Prologue to a Political Theory of the Steady State*. San Francisco: W.H. Freeman, 1977.

313 Patterson, W., *Nuclear Power*. Harmondsworth: Penguin, 1976.

314 – *The Fissible Society*. London: Earth Resources Research, 1977. (E.R.R: 40, James Street, London W1M 5HS, England.)

315 Patterson, W. and R. Griffin, *Fluidized Bed Energy Technology: Coming to a Boil*. New York: INFORM, 1978 (see also the review in *Soft Energy Notes*, 1 (October 1978), 75–8).

316 Peat, D., *The Nuclear Book*. Ottawa: Deneua and Greenberg, 1979.

317 Penner, S.S. and L. Icerman, *Energy*. 3 vols. Reading, Mass.: Addison-Wesley, 1974, 1975, 1976.

318 Penney, K., 'Economics for a Post-industrial Society,' *Ecologist*, 9 (September/October, 1979), 200–8.

319 Peters, R. *Municipal Potential for Energy Conservation and the Use of Renewable Energy Sources*. Toronto: National Survival Institute, October 1979 (draft).

320 – 'Strategy for a Consumer-Regulated Economy and Energy System,' *Conserver Society Notes*, (winter 1979).

321 – *Techniques of Community Energy Conservation*. A Report for the Consumer Interest Study Group, Consumer and Corporate Affairs, Ottawa. Ottawa, July 1977.

322 Phillips, O., *The Last Chance Energy Book*, Baltimore: Johns Hopkins University Press, 1979.

323 Pimlott, D., et al., *Oil under the Ice: Off Shore Drilling in the Canadian Arctic*. Ottawa: Canadian Arctic Resources Commission, 1976.

324 Piraeges, D.C. (ed.), *The Sustainable Society: Implications for Limits to Growth*. New York: Praeger, 1977.

325 Platts, J., 'Electrical Load Management: The British Experience,' *Spectrum*, 16 (1979), Part I: 39–42 (February), Part II: 66–68 (April).

326 Pohl, R.O., 'Health Effects of Radon-222 from Uranium Mining,' *Search*, 7 (August 1976) 345–50.

327 Pratt, L., *The Tar Sands: Syncrude and the Politics of Oil*. Edmonton: Hurtig, 1976.

328 Puttagunta, V.R., *Temperature Distribution of the Energy Consumed as Heat in Canada*. Atomic Energy of Canada Limited, October 1975 (AECL-5235).

329 Quinn, J. and J. Ohioe, *A Compendium of Decentralized Studies, Programs and Projects* (draft). Washington, DC: U.S. Department of Energy, Office of Solar, Geothermal and Electric Systems, May 14, 1979. (OSGES: 600 E. St. NW, Washington DC, 20545).

330 Randall, M., '*Energy Policy Planning: Towards a Critical Perspective,*' *Alternatives*, 6 (1977), 13–26.

331 Ranger Uranium Environmental Inquiry, *First Report*. Presiding Commissioner: Justice R.W. Fox. Canberra: Australian Government Publishing Service, 1976.

332 Reuyl, J.S., et al., *Solar Energy in America's Future*, 2nd ed. Menlo Park, Calif.: Stanford Research Institute, 1977.

333 Ridgway, J., *The Last Play: The Struggle to Monopolize the World's Energy Resources*. New York: Mentor, 1973.

334 Ridgway, J. and B. Conner, *New Energy: Understanding the Crisis and a Guide to an Alternative Energy System*. Boston: Beacon Press, 1975.

335 Roberts, A., 'The Politics of Nuclear Power,' *Arena*, 41 (1976), 22–47.

336 Robertson, D.S. and Associates Limited, *A Brief Review of Uranium Supply and Demand*. Toronto, December 8, 1977.

337 Robertson, J., *The Sane Alternative: Signposts to a Self-fulfilling Future*. London: James Robertson, 1978 (available from 7 St. Ann's Villas, London W11 4RU).

338 Robinson, J., *Canadian Energy Futures: An Investigation of Alternative Energy Scenarios, 1974–2025*. A Report for the Work Group on Canadian Energy Policy, Faculty of Environmental Studies, York University. Toronto, 1977.

339 Rockingham, A.P., *The Impact of New Energy Sources on Electric Utilities*. A report prepared for the Royal Commission on Electric Power Planning, Toronto, May 1979.

340 Rocks, L. and R.P. Runyon, *The Energy Crisis*. New York: Crown Publishers, 1972.

341 Rogers, C., *On Becoming a Person*. Boston: Houghton Mifflin, 1961.

342 Romanos, M., *Energy Conservation through Land-Use Planning: A Framework for Research*. Urbana-Champaign, Ill.: Department of Urban and Regional Planning, November 1976.

343 Rosenberg, D., 'James Bay Up-date.' *Alternatives*, 4 (1974), 4–10.

344 Ross, D., *Energy from the Waves*. Oxford: Pergamon, 1979.

345 Ross, D.W., and Associates, *What Are the 'Reasonably Foreseeable' Requirements for Energy for All Uses in Ontario; with Specific Reference to Electricity and to the Next Fifteen Years to the End of the 1980's?* A presentation to the Select Committee of the Legislature Investigating Ontario Hydro, 1976.

346 Ross, M.H. and R.H. Williams, *Energy and Economic Growth.* A paper prepared for the subcomittee on Energy of the Joint Economic Committee, Congress of the United States. Washington, DC: U.S. Government Printing Office, 1977.

347 – 'The Potential for Fuel Conservation,' *Technology Review*, 79 (No. 4, 1977), 48–58.

348 Ross, W.A., 'Energy Powers for Canada,' *Alternatives*, 7, 1 (fall 1977), 4–7.

349 Roszak, T., *The Making of a Counter-Culture.* New York: Anchor, 1969.

350 – *Person/Planet: The Creative Disintegration of Industrial Society.* Garden City, NY: Doubleday, 1978.

351 Rotblat, J., 'The Risks for Radiation Workers,' *Bulletin of the Atomic Scientists*, 34 (September 1978), 41–6.

352 Routley, V. and R., 'Social Theories, Self-management and Environmental Problems,' in D. Mannison, M.A. McRobbie, R. Routley (eds.), *Environmental Philosophy.* Canberra: Research School of Social Sciences, Australian National University, 1979.

353 Rowland, W., *Fueling Canada's Future.* Toronto: Macmillan, 1974.

354 Royal Commission on Electric Power Planning, *A Race against Time: Interim Report on Nuclear Power in Ontario.* Chairman: Dr Arthur Porter. Toronto: Queen's Printer, 1978.

355 – *Our Energy Options.* Toronto: Queen's Printer, 1978.

356 Royal Commission on Environmental Pollution, *Sixth Report: Nuclear Power and the Environment.* Chairman: Sir Brian Flowers. London: Her Majesty's Stationery Office, 1976.

357 Ruedisili, L.C. and M.W. Firebaugh, *Perspectives on Energy.* Oxford: Oxford University Press, 1975.

358 Ryan, C., *The Choices in the Next Energy and Social Revolution.* Houston, Texas: Woodlands Conference, 1977 (available from University of Texas, Houston, Texas, 77004).

359 Saltzman, F.A. (ed.), *Energy, Technology and Global Policy.* Santa Barbara, Calif.: Center for the Study of Democratic Institutions, 1977.

360 Sampson A., *The Seven Sisters: The Great Oil Companies and the World They Made.* New York: Viking, 1975.

361 Satin, M., *New Age Politics: Healing Self and Society: The Emerging New Alternative to Marxism and Liberalism.* West Vancouver, BC: Whitecap Books, 1978.

362 Saskatchewan Legislature, *Report of the Cluff Lake Board of Inquiry*, Mr Justice Bayda, Chairman, Saskatoon, May 1978.
363 Sayre, K. (ed.), *Values in the Electric Power Industry*. Notre Dame, Ind.: University of Notre Dame Press, 1977.
364 Schipper, L., 'Raising the Productivity of Energy Utilization,' *Annual Reviews of Energy*, 1 (1976), 455–517.
365 Schipper, L. and A.J. Lichtenberg, 'Efficient Energy Use and Well-being – The Swedish Example,' *Science*, 194 (December 1976), 1001–13.
366 Schipper, L. and J. Darmstadter, 'Energy Conservation,' *Technology Review.*, 80 (January 1978), 41–50.
367 Schumacher, E.F. *Small Is Beautiful: A Study of Economics as if People Mattered*. London: Abacus, 1974.
368 Schumacher, E.F. and P.N. Gillingham, *Good Work*. New York: Harper & Row, 1979.
369 Schurgin, A.S. and T.C. Hollocher, 'Radiation Induced Lung Cancers among Uranium Miners,' in [427].
370 Science Council of Canada, *Canada's Energy Opportunities*. Ottawa: Queen's Printer, March 1975.
371 – *Nuclear Dialogue*. Ottawa, 1976.
372 – *Canada as a Conserver Society: Resource Uncertainties and the Need for New Technologies*. Report 27, Dr U. Franklin, chairperson, Ottawa: Queen's Printer, 1977.
373 – *Uncertain Prospects: Canadian Manufacturing Industry 1971–1977*. Ottawa: Queen's Printer, October 1977.
374 – *Roads to Energy Self-reliance: The Necessary National Demonstrations*. Ottawa: Queen's Printer, June 1979.
375 Scientific American, *Energy and Power*. San Francisco: W.H. Freeman, 1971.
376 Scitovsky, T., *The Joyless Economy*. Oxford: Oxford University Press, 1976.
377 Senate Special Committee on Science Policy, *A Science Policy for Canada*, I. Ottawa: Queen's Printer, 1970 (N.B. vol. II, 1972; vol. III, 1973).
378 Sevc, J., E. Kunz, and V. Placek, 'Lung Cancer in Uranium Miners and Long-Term Exposure to Radon Daughter Products,' *Health Physics*, 30 (1976), 433–7.
379 Slater, P., *Earth Walk*. New York: Anchor, 1974.
380 Smil, V., *China's Energy: Achievements, Problems, Prospects*. New York: Praeger, 1976.
381 – 'Intermediate Energy Technology in China,' *Bulletin of the Atomic Scientists*, 33 (February 1977), 25–31.
382 Smith, P., *Brinco: The Story of Churchill Falls*. Toronto: McClelland & Stewart, 1975.

383 s.n.c. Consultants, *Hearst Woodwaste Energy Study*. Toronto, 1976. (See also the summary report by Acres Shawinigan Ltd., January 1979.),

384 Socolow, R.H., 'The Coming Age of Conservation,' Palo Alto., Calif.: *Annual Reviews of Energy*, 2 (1977), 239–90.

385 Solar Lobby, *Blueprint for a Solar America*. Washington, DC, January 1979.

386 Solomon, L., *The Conserver Solution: A Blueprint for the Conserver Society*. A Project of the Pollution Probe Foundation. Toronto: Doubleday Canada, 1978.

387 Sørensen, B., *Renewable Energy*. Copenhagen: Academic Press, Niels Bohr Institute, University of Copenhagen, 1979.

388 Speed, T.P., *Negligible Probabilities and Nuclear Reactor Safety: Another Mis-use of Probability?* Western Australia: Department of Mathematics, University of Western Australia, May 1977.

389 Spence, J.A. and G.C. Spence, *Ecological Considerations of the James Bay Project*. A report prepared for the United Nations Conference on the Human Environment, Montreal, April 1972.

390 Speth, J.G., A.R. Tamplin, and T.B. Cochran, 'Plutonium Recycle: The Fateful Step,' *Bulletin of the Atomic Scientists*, 33 (November 1974), 15–22.

391 Starr, C., *Electricity Needs to the Year 2000 as Related to Employment, Income and Conservation*. A Report presented to the subcommittee on Energy Research Development and Demonstration, House Committee on Science and Technology, United States. Washington, DC, 1976.

392 – 'Social Benefit versus Technological Risk,' *Science*, 165 (1969), 1232.

393 Steadman, P., *Energy, Environment and Building*. Cambridge, Mass.: MIT Press, 1976.

394 Stein, B.A., *Size, Efficiency and Community Enterprise*. Cambridge, Mass.: Center for Community Economic Development, 1974.

395 Stein, R.G., *Architecture and Energy*. New York: Anchor, 1978.

396 Stewart, G. and C. Starrs, *Gone Today, Here Tomorrow*. A working paper prepared for the Committee on Government Productivity, Ontario, Canada. Toronto: Queen's Printer, 1972.

397 Stiefel, M., 'Soft and Hard Energy Paths: The Roads Not Taken?,' *Technology Review*, 81 (October 1979), 56–66.

398 Stobaugh, R., G. Yurgin, et al., *Energy Future: Report on the Energy Project at the Harvard Business School*. New York: Random House, 1979.

399 Stokes, B., *Local Response to Global Problems: A Key to Meeting Basic Human Needs*. Washington, DC: Worldwatch Institute, 1978.

400 Sullivan, G., *Wind Power for Your Home*. New York: Cornerstone Library, 1978.

401 Surrey, A.J., et al., *Democratic Decision-Making for Energy and the Environ-ment*. London: Macmillan, 1978.
402 Swain, H., R. Overend, and T.A. Ledwell, *Canadian Renewable Energy Prospects*. Energy, Mines and Resources Canada: Renewable Energy Resources Branch. Ottawa, undated.
403 Swartman, R.K., 'Alternative Power Generation Technologies,' in 355.
404 Swedish Secretariat for Future Studies, *Energy in Transition: A Report on Energy Policy and Future Options*. Uddevalla, Sweden: Swedish Institute, 1977.
405 Sykes, P. *Sellout*. Edmonton: Hurtig, 1973.
406 Tamblyn, R., *The Case for Thermal Storage*. Toronto: Engineering Interface, July 1975.
407 Tamplin, A.R., *The Biological Affects of Radiation: Ten Times Worse than Estimated*. Washington, DC: Natural Resources Defence Council, August 1977.
408 Tartaryn, L., *Dying for a Living*. Ottawa: Deneua and Greenberg, 1979.
409 Taylor, V., *Energy: The Easy Path*. Cambridge, Mass.: Union of Concerned Scientists, January 1979.
410 – *The Easy Path Energy Plan*. Cambridge, Mass.: Union of Concerned Scientists, September 1979.
411 Templin, R.J. 'Wind Energy.' A paper presented at the Third Canadian National Energy Forum, Halifax, April 5, 1977.
412 Tester F. (ed.), *Proceedings, First Canadian Conference on Social Impact Assessment*, Banff, May 1979, in press.
413 Thayer, R., *An End to Hierarchy, an End to Competition*. New York: New Viewpoint, 1973.
414 The Public Resource Center, *The Davis Experiment: One City's Plan to Save Energy*. Washington, DC: Public Resource Center, 1977.
415 Theobald, R., *Habit and Habitat*. Englewood Cliffs, NJ: Prentice-Hall, 1972.
416 Theobold, R. and J.M. Scott, *Tegs 1994*. Chicago: Swallow Press, 1972.
417 Thompson, D. and H. Boerma, *Saskatchewan Energy Use and Renewable Energy Supply: Three Scenarios for 2025*. Saskatoon: Saskatchewan Research Council, March 1979.
418 Thompson, W.I., *Evil and World Order*. New York: Harper Colophon, 1976.
419 Toombs, R. 'Petroleum Supply and Demand in Canada, 1977–1990,' a refer-ence paper prepared in the Energy Policy Section, Department of Energy, Mines and Resources, Canada, Ottawa, for the Industrial Energy Conserva-tion Seminar, Ottawa, November 23, 1977.
420 Torgerson, D., 'From Dream to Nightmare: The Historical Origins of Canada's Nuclear Electric Future,' *Alternatives*, 7, 1 (fall 1977), 8–17, 30, 63–4.

421 Torrie, R., 'People ... and Other Hazards of Nuclear Technology,' *Alternatives*, 7, 1 (fall 1977), 18–25.

422 – 'Reactor Safety: Too Many Unanswered Questions,' *Alternatives*, 7, 1 (fall 1977), 34–43.

423 Torrie, R. and G. Edwards, *Summary Argument to the Royal Commission on Electric Power Planning*, Montreal: Canadian Coalition for Nuclear Responsibility, 1978.

424 Tribe, L., 'Technology Assessment and the Fourth Discontinuity: The Limits of Instrumental Rationality,' *Southern California Law Review*, 46 (1973), 617–60.

425 Uffen, R.J., *The Disposal of Ontario's Used Nuclear Fuel.* A status report on alternative proposals for the storage, reprocessing and ultimate disposal of used fuel from CANDU Nuclear Reactors for Ontario Hydro. Kingston and Toronto, 1978.

426 Unger, R., *Knowledge and Politics.* New York: Free Press, 1975.

427 Union of Concerned Statistics, *The Nuclear Fuel Cycle.* Cambridge Mass.: MIT Press, 1975.

428 – *The Risk of Nuclear Power Reactors: A Review of the NRC Reactor Safety Study, WASH-1400.* Study Director: Dr H. Kendall. Cambridge, Mass.: Union of Concerned Scientists, 1977.

429 United Nations, *Ionizing Radiation: Levels and Effects.* 2 vols. A report of the United Nations Scientific Committee on the Effects of Atomic Radiation. New York: United Nations, 1972.

430 U.S. Congress, *Hearings on Radiation Exposure of Uranium Miners.* Proceedings, Joint Committee on Atomic Energy, Sub-Committee on Research, Development and Radiation, 90th Congress of the United States. Washington: U.S. Government Printing Office, 1967.

431 U.S. Council on Environmental Quality, *The Good News About Energy.* Washington, DC, 1979.

432 – *Solar Energy: Prospects and Promises.* Washington, 1979.

433 U.S. Health, Education and Welfare, *Inter-Agency Task Force on Ionizing Radiation Report* (Chairman: P. Libassi). Washington, DC: U.S. Government Printing Office, March 1976.

434 U.S. Senate, *Alternative Long-Range Energy Strategy.* Joint hearings of the Senate, Small Business and Interior Committees, Washington: U.S. Government Printing Office, 1977.

435 Valaskakis, K., P.S. Sindell, J.G. Smith, and I. Fitzpatrick-Martin, *The Conserver Society: A Workable Alternative for the Future.* New York: Harper & Row, 1979.

436 Van Hulst, R., 'The Dangers of Nuclear Power Planning,' in *221*.

437 Verna, B.J., 'Radioactive Maintenance,' *Nuclear News*, 18 (September 1975), 54–5.

438 Vickers, G., *Freedom in a Rocking Boat*. London: Pelican, 1972.

439 – 'The Weakness of Western Culture,' *Futures*, (December 1977), 457–73.

440 – *Value Systems and the Social Process*. London: Tavistock, 1968.

441 Victor, P.A., 'Economics and the Challenge of Environmental Issues,' in *221*.

442 von Hippel, F. and R.H. Williams, 'Energy Waste and Nuclear Power Growth,' *Bulletin of the Atomic Scientists*, 32 (December 1976), 18.

443 Wallace, C., 'Brazilian National Alcohol Program,' *Soft Energy Notes*, VII (July 1979), 60–3.

444 Wallenstein, A.R., *Barriers and Incentives to Solar Energy Development: An Analysis of Legal and Institutional Issues in the Northeast*. Northeast Solar Energy Center, December 1978.

445 Watkins, C. and M. Walker, (eds.), *Oil in the Seventies: Essays on Energy Policy*. Vancouver, BC: Fraser Institute, 1977.

446 Wayman, M., 'New Opportunities for Fuel From Biological Processes,' *Chemistry in Canada*, 29 (November 1977), 26–9.

447 Webb, R.E., *The Accident Hazards of Nuclear Power Plants*. Amherst, Mass.: University of Massachusetts Press, 1976.

448 Weinberg, A., 'Social Implications of Nuclear Energy,' in 121.

449 – 'Social Institutions and Nuclear Power,' *Science*, 28, 5 (1972).

450 – 'The Many Dimensions of Scientific Responsibility,' *Bulletin of the Atomic Scientists*, 32 (November 1976), 21–4.

451 Weisskopf, V., *Alienation and Economics*. New York: Delta, 1971.

452 White, D., et al., *Seeds for Change: Creatively Confronting the Energy Crisis*. Melbourne, Australia: Patchwork Press and Conservation Council of Victoria, 1978.

453 Whiting, M., 'Industry Saves Energy: Progress Report, 1977,' *Annual Reviews of Energy*, 3 (1978), 181–200.

454 Widmer, T.F. and E.P. Gyftopoulos, 'Energy Conservation and a Healthy Economy,' *Technology Review*, 79, 7 (June 1977), 30–41.

455 Wiggins, E.J., *Prospects for Solar and Wind Energy Utilization in Alberta*. A Report for the Department of Energy and National Resources, Government of Alberta. Edmonton: Energy and Natural Resources, December 1978.

456 Williams, R.H., 'Industrial Co-generation,' *Annual Reviews of Energy*, 3 (1978), 313–56.

457 – *Toward a Solar Civilization*. Cambridge, Mass.: MIT Press, 1978.

458 Willrich, M., *Energy and World Politics*. New York: Free Press, 1975.
459 Wilson, C.L., *Energy: Global Prospects, 1985–2000*. New York: McGraw-Hill, 1977.
460 Woollard, R.F. and E.R. Young, *Health Dangers of the Nuclear Fuel Chain and Ionizing Radiation: A Bibliography/Literature Review*, Vancouver: British Columbia Medical Association, May 1979 (available from Academy of Medicine Building, 1807 West 10th Avenue, Vancouver V6J 2A9).
461 Yannacone, Z.J. Jr., *Energy Crisis: Danger and Opportunity*. New York West: West Publishing, 1974.

Contacts in the alternative energy field

Much of the available information on energy resources, technologies, and policy, especially that concerned with the existing 'conventional' fossil/nuclear approach, is easily obtainable from Energy Mines and Resources Canada, the Ontario Ministry of Energy, Ontario Hydro, and Atomic Energy of Canada Limited. The first two of these institutions also have groups actively studying conservation and renewable energy technologies and can provide technical information.

For comprehensive information on 'alternative' energy policies and technologies, however, the reader must make contact with the public interest groups in the field. Many of them are informal, local organizations, and there are far too many of them to list here. But we do want to provide the reader with an entrance to the network. First, then, we list some of the most important Canadian groups and some Canadian publications that are important sources of information on energy alternatives. We also list some, though by no means all, U.S. groups active in the same area and append three non-Canadian energy newsletters as initial contacts for readers. With these introductions, the reader can begin exploring the rich and fascinating unofficial world of alternative approaches. We understand that, in connection with Energy Probe, Lawrence Solomon's *Energy Shock* and Jan Mamorek's *Guide to the Canadian Energy Crisis* will appear shortly from Doubleday Press; David Brooks' *Zero Energy Growth for Canada* is in press at McClelland & Stewart.

Notably absent from this brief list are academic and private consultant groups and those concerned with 'third world' or 'developing nations' policy. All of them make important contributions to energy policy; however, the interested reader will quickly learn of them through the other groups and publications listed below and by consulting the bibliography.

Also absent is any mention of the many excellent audiovisual materials now becoming available in the energy area. Readers are advised to contact especially the Canadian Coalition for Nuclear Responsibility, the Development Education Centre,

Toronto (121 Avenue Road, M5R 2G3), the National Film Board of Canada, and especially the latter's 'Challenge for Change' series and the resource booklet to accompany the film *The New Alchemists*.

CANADA

Canadian Arctic Resources Commission: 46 Elgin Street, Room 11, Ottawa, Ontario, K1P 5K6

Canadian Coalition for Nuclear Responsibility: 848 Somerset Street W., Ottawa, Ontario, K1R 6R7; 2030 Mackay Rue, Montréal, Québec, H3G 2J1.

Energy Probe: 53–4 Queen Street, Ottawa, Ontario, K1P 5C5

Friends of the Earth (Canada): P.O. Box 569, Station B, Ottawa, Ontario, K1P 5P7

National Survival Institute (Canada): 229 College Street, 3rd Floor, Toronto, Ontario, M5T 1R4

Ontario Public Interest Research Group: Trent University, Peterborough, Ontario, K9J 7B8

Pollution Probe: 43 Queen's Park Crescent E., Toronto, Ontario, M5S 2C3

Solar Energy Society of Canada Inc.: Suite 303, 870 Cambridge Street, Winnipeg, Manitoba, R3M 3H5

Alternatives: Conserver Society Notes: Trent University, Peterborough, Ontario, K9J 7B8

Canadian Renewable Energy News: P.O. Box 4869, Station E, Ottawa, Ontario, K1S 5B4

Probe Post: 43 Queen's Park Crescent E., Toronto, Ontario, M5S 2C3

Soft Energy Notes: Journal of the International Project for Soft Energy Paths: 124 Spear Street, San Francisco, California 94105 (Contact David Brooks, Energy Probe, 53–4 Queen Street, Ottawa, Ontario, K1P 5C5)

The Conserver: Conservation and Renewable Energy Branch, Energy, Mines and Resources, 580 Booth Street, Ottawa, Ontario, K1A 0E4

UNITED STATES

Environmental Policy Institute: 317 Pennsylvania Avenue S.E., Washington, D.C. 20003

Environmentalists for Full Employment: 1785 Massachusetts Avenue N.W., Washington, D.C.

Friends of the Earth (International): 124 Spear Street, San Francisco, California 94105

INFORM: 25 Broad Street, New York, New York 10004

Institute for Ecological Policies: 9208 Christopher Street, Fairfax, Virginia 22031

Institute for Local Self Reliance, 1717 18th Street N.W., Washington, D.C. 20009

Northeast Appropriate Energy Technology Network: P.O. Box 548, Greenfield,
    Massachusetts 01301

Sierra Club: 530 Bush Street, San Francisco, California 94108

Solar Lobby: 1028 Connecticut Avenue N.W., Washington, D.C. 20036

The Union of Concerned Scientists: 1208 Massachusetts Avenue, Cambridge,
    Massachusetts 02238

TRANET, Transnational Network for Appropriate/Alternative Technologies:
    P.O. Box 567, Rangeley, Maine 04970

EUROPE

Turning Point: The Grove, 10 New Road, Ironbridge, Telford, Salop TF8 7AU,
    England

World Information Service on Energy: 2c Weteringsplantsoen 9, 1017 ZD
    Amsterdam, Netherlands